MW01485278

The
**PRINCETON
FIELD GUIDE** *to*

MESOZOIC
PLANTS

The
PRINCETON
FIELD GUIDE *to*
MESOZOIC
PLANTS

NAN CRYSTAL ARENS

Princeton University Press
Princeton and Oxford

Published by Princeton University Press
41 William Street, Princeton, New Jersey 08540
99 Banbury Road, Oxford OX2 6JX
press.princeton.edu

GPSR Authorized Representative: Easy Access System Europe—
Mustamäe tee 50, 10621 Tallinn, Estonia, gpsr.requests@easproject.com

Copyright © 2025 by UniPress Books Limited
Illustrations copyright © 2025 by Sante Mazzei
www.unipressbooks.com

No part of this publication may be reproduced, stored in a
retrieval system, or transmitted, in any form or by any means,
electronic, mechanical, photocopying, recording, or otherwise,
without the prior permission of the publishers. Requests for
permission to reproduce material from this work should be sent to
permissions@press.princeton.edu.

The right of Sante Mazzei to be identified as Illustrator of this Work has
been asserted in accordance with the Copyright, Designs and Patents
Act 1988.

All Rights Reserved

ISBN 978-0-691-27243-6
Ebook ISBN 978-0-691-27249-8

Library of Congress Control Number: 2025931194
British Library Cataloging-in-Publication Data is available

Commissioning Editor: Claire Collins
Project Manager: David Price-Goodfellow
Text Design: Namrita Price-Goodfellow
Copyeditor: Jenny Dixon
Jacket Design: Shane O'Dwyer

Illustrator: Sante Mazzei

This book has been composed in LTC Goudy Oldstyle Pro (main text)
and Acumin Pro (headings and captions)

Drawing previous page: *Antarcticycas schopfii*
Cover images: top, *Taxus guyangensis*; right, *Pseudoctenis lanei*;
bottom, *Marattia intermedia*

Cover art by Sante Mazzei. Background shadows courtesy of
Shutterstock/TanyaFox

Printed in Malaysia

10 9 8 7 6 5 4 3 2 1

CONTENTS

PREFACE

Aliens Among Us is a long-standing trope of science fiction. Strange creatures from another world invade, take over the planet, and bend earthlings to their will. In the most frightening versions of this story, the hapless humans are unaware of or perhaps beguiled by their alien overlords. But that is science *fiction*. In science fact, we have plants. They did not come from another world. Instead, they evolved from freshwater green algae alongside our fishy ancestors and pioneered, then transformed, life on land. In fact, land plants sculpted the atmosphere and terrestrial environment to make it habitable for animals like us. Yet plants are so alien that some people question whether they are life-forms at all. To the casual observer, plants may seem inanimate. They move too slowly for us to perceive, yet shift your houseplant away from the window and it will turn back toward the light like the monstrous statue that moves only when you look away. Plants "see" without eyes, "smell" without noses, "breathe" without lungs, move fluid around their bodies without hearts, and communicate with each other and with their animal partners in the complex chemical language of scent. Plants are simultaneously wondrously alien and so intimately connected to us that we literally could not live without them. They freshen the air that we breathe, help the rain to fall, create and stabilize the soil, and make food enough for everyone from sunlight, air, and water.

Yet when the fantastic story of Earth's history is told, plants—if they appear at all—are usually the backdrop to the animal action. This work brings the little (and big) green aliens to center stage and tells the remarkable story of how they created, then transformed, the Mesozoic Earth and fashioned the world we inhabit today.

ACKNOWLEDGMENTS

Many thanks to Claire Collins at Unipress, who extended the most surprising invitation, on behalf of Princeton University Press, to create this book. Thanks to David and Namrita Price-Goodfellow for patient answers to the naive questions of a first-time commercial author, unfailingly good counsel, precise and beautiful design, and for keeping the project on track and rolling smoothly through the Mesozoic countryside. Thanks to Jennifer Dixon for thoughtful copyediting and to the many other hands working behind the scenes. Gratitude to Sarah Kirk, David Finkelstein, and David Kendrick for helping make space in my academic life to complete this project, and to my students who waited more patiently than usual for work to be returned. Special thanks to David, Laurel, Rose, Diana, and Rikki Tikki Tavi the Sheltie for providing boundless support and encouragement. Tracie the Corgi probably helped, too, but it's hard to pinpoint exactly how—that is the nature of corgis. And kudos to Sante Mazzei, who embraced the challenge of bringing these aliens to life.

Somewhere in the shadow of the great trees, far away from the places where fossils usually form, the Mesozoic flowering plant revolution began as the upstart angiosperm lineage adapted to damp, dark, and disturbed understory habitats. The nutritious leaves and delicious fruit produced by low-growing flowering plants changed the course of dinosaur evolution, making room for ground-foragers like *Minmi paravertebra*, and providing abundant energy for up-and-coming mammals like *Spinolestes*.

INTRODUCTION

At the end of the Paleozoic era, 252 million years ago, Earth baked in a climatic hothouse. Over the course of two million years, volcanoes covered nearly 5 million square kilometers (2 million square miles) of what is now central Siberia, erupting lava that solidified into basalt rock up to 3 km (almost 2 mi) thick. Carbon dioxide belching from the volcanic fissures boosted the already toasty temperatures so much that the sea surface approached 30°C–32°C (86°F–90°F), even near the poles. Warm polar water blocked the currents that normally refreshed the depths, and over time the ocean grew stagnant and devoid of oxygen. Marine animals suffocated; microbes that thrived without oxygen feasted on the carcasses. Their toxic waste gases, including methane and hydrogen sulfide, bubbled into the atmosphere and spread the death onto land with devastating consequences. The Permian extinction—the "Great Dying," as this event is known—killed roughly 81% of marine species and about 70% of animals on land. The plants, however, barely noticed. Estimates of land plant extinction, although complicated by a spotty fossil record, may have been as little as 20%. Ferns and cycads lost some species, while conifers and ginkgos passed, little-changed, into the Mesozoic, the era of geological history that includes the Triassic (252–201 million years ago), Jurassic (201–145 million years ago), and Cretaceous (145–66 million years ago).

As the Earth slowly recovered, the survivors grew to dominate Triassic forests. Ferns and cycads rebounded, evolving new diversity surpassing even that lost during the extinction. New lineages appeared, with armored stems, spiky leaves, and showy reproductive structures for attracting pollinators. And in the shadowy understory, an entirely new group emerged. Its seeds germinated best in ground churned up by the passage of lumbering dinosaurs. A sweet scent wafted from its delicate petals, promising a nectar treat to passing insects. Its glossy, deep-green leaves had veins connected in a network that could quickly move water to the sites of photosynthesis if a speckle of sun fell on them.

This new group, the flowering plants—angiosperms—probably originated in the Triassic or Jurassic but did not appear in the fossil record until early in the Cretaceous, probably because they grew far from the environments where fossils usually form. Then they burst into the sunny streamsides and began their conquest of the globe. In the Early Cretaceous, flowering plants were rare, but by about 100 million years ago, they were widespread. New species arose quickly and escaped the tropical latitudes, advancing toward both poles. By the end of the Cretaceous, conifers, cycads, ginkgos, and a variety of other seed plants had become less common, relegated to specialized habitats or standing solitary within forests of flowers. Flowering plants on the other hand, and the ferns that diversified with them, seized control of land worldwide, reshaping entire ecosystems and the animals that occupied them. The delicate foliage and nectar-rich flowers of angiosperms provided food for newly evolved insects like bees and butterflies. Angiosperm shrubs and herbs transformed herbivorous dinosaurs from long-necked giants adapted for foraging in the treetops to ceratopsians and ankylosaurs that browsed close to the ground. Nutritious and delicious angiosperm fruit and seeds sustained calorie-hungry mammals that scurried in the shadows. And when another great outpouring of magma and a 10 km (over 6 mi) asteroid devastated Earth at the end of the Cretaceous, these new plants barely noticed. Primed for disturbance, the angiosperms bounced back to help the earliest Cenozoic terrestrial ecosystems recover surprisingly quickly.

The story of Mesozoic plants is above all one of transformation. The sturdy survivors of the end-Permian hothouse thrived by growing slowly and investing in defense. Then flowering plants changed the rules; they grew fast and cheap, prioritizing reproduction and adaptation to a world where flood, wildfire, and giant herbivores might destroy them at any moment. In so doing, flowering plants grabbed a roothold in previously empty spots on the landscape. Their grow-cheap, reproduce-fast lifestyle honed by disturbance eventually crowded their neighbors out of even the most desirable locations. The "Flowering Plant Revolution" of the Mesozoic did not begin with an extinction like the major milestones in animal history. Instead, a novel combination of traits allowed flowers to first infiltrate, then dominate, and then completely remake nearly every corner of the terrestrial world.

HISTORY OF DISCOVERY AND RESEARCH

The scientific study of fossil plants began with a chance meeting in Paris. As the third son of an aristocratic Bohemian family, Kaspar Maria von Sternberg had been expected to find a career in the Church. But his heart belonged to plants. During a stint as a Church diplomat in Paris (1804–5), Sternberg drifted into the circle of the maverick French paleontologist George Cuvier. Cuvier had famously proposed that the elephant-like skeletons he had unearthed near Paris were the remains of extinct animals, related to but different from any contemporary elephant. The idea was controversial because it challenged the Church's assertion that all species were created perfect and unchanging. Through Cuvier, Sternberg met French botanist Adolphe-Théodore Brongniart, who had collected a few plant fossils from nearby mines and considered their relationship to living plants. Brongniart wondered whether the slabs of gray shale that he collected from coal mines across Europe contained the remains of plants that once grew in the French countryside. Long ago, he speculated, the region had been a tropical wetland crisscrossed by streams that entombed fallen plants in sediment. Sternberg's mind wandered to the coal mines in the valley below his home in today's Czech Republic; were they also full of the enigmatic fossils? With his brother's blessing—and financial backing—Sternberg abandoned his career in the Church and established a botanical garden near his home at Březina Castle. Exploring the spoil piles of the coal mines near his estate, Sternberg found fossils that resembled those Brongniart had shown him. Sternberg meticulously described and drew each scrap, using the same botanical eye that he applied to living plants. He realized that the impressions in stone had once been living plants that were unlike any alive at the time. With that insight, Sternberg and Brongniart laid the foundations for the science of paleobotany—the study of the plants of the deep past, discoverable as fossils preserved in stone. The paleobotanists who followed them would be both experts in botany—fluent in the biology, evolutionary relationships, and ecology of living plants—*and* adept geologists, ready to trek into remote areas and split rock to reveal the secrets within.

Revealing the Dinosaur Forest

Although the first dinosaur fossil, *Megalosaurus*, was unearthed from the Middle Jurassic rocks of England in 1677, until the early nineteenth century, almost nothing was known about the forest in which that two-legged carnivore hunted. In 1813, William Smith, the English geologist known for first recognizing that layers in sandstone, shale, and coal could be traced across Europe by their particular combination of fossils, visited northeastern Yorkshire to study coal mines. During the visit, Smith encountered John Williamson, the founding curator of the Scarborough Museum, who led him to the seaside cliffs where he found fossil plants. Smith was intrigued and published a detailed geological map of the area in 1821. Recognizing that there was more work to do, Smith and his nephew, John Phillips, moved to Scarborough in 1824, making it their base for an extensive study of the region, and in 1829 Phillips published the first detailed drawings of the plants collected from the cliffs. Smith and Phillips knew from the abundance of conifers that the fossils were of Mesozoic age. The report caught the attention of Adolphe-Théodore Brongniart, who requested samples and included 22 species of Yorkshire Jurassic plants in later volumes of his *Histoire des Végétaux Fossile*.

Local collectors including William Bean, John Williamson's son William, and John Phillips returned to the cliffs regularly and opened dozens of new collecting sites. Together, they supplied a steady stream of fossils to geologist William Hutton, who shared some with his colleague John Lindley at University College, London. Between 1831 and 1837, Lindley and Hutton published descriptions and illustrations of several species in the first detailed account of fossil plants in Britain—*Fossil*

Flora of Great Britain. However, it was really William Williamson who had written the descriptions and drawn the illustrations published in the *Fossil Flora*, a fact Lindley only discovered when Williamson enrolled as a student at University College in 1840.

Although the younger Williamson earned his living as a physician, fossils were his passion. He built and displayed an extensive collection of fossils at the Museum of Natural History in Manchester and continued to collect fossils along the coastal cliffs of Yorkshire, discovering many new sites. Williamson lectured passionately about Yorkshire's fossils, but despite his enthusiasm and substantial collection, the Jurassic plants of Yorkshire failed to catch the eye of science. In 1886, the young Albert Seward visited the charismatic Williamson and, after spending a year poring over the Yorkshire collections, abandoned his plan for a career in the Church to study Mesozoic plants. Unlike Williamson, Seward was not particularly interested in collecting fossils in the field. Travel was time-consuming, academic tasks kept professors tied to their universities, and many shied away from the sweaty, grimy labor of excavating fossils. Europe's museums were already filled with unstudied fossils, Seward reasoned, and he could build a successful career studying those. By 1900 he had cataloged all the Jurassic plants in the British Museum of Natural History. But despite his encyclopedic knowledge of Jurassic plants, Seward's perspective echoed that of Sternberg and Brongniart: Fossils were "pictures on the rock," like pressed specimens from the ancient past. Seward viewed fossil plants much as a botanist might, describing their features, naming and classifying them, and considering their evolutionary relationships, but he could not imagine the forests or even, in some cases, the plants from their parts.

Plants tend to fall apart before they become fossils. Pollen and seeds disperse and may be preserved kilometers away from the parent plant. Leaves fall off, and woody trunks may float far from the forest where they grew before being buried and preserved. When paleobotanists find fossils of isolated plant parts, it can be difficult to reunite them to reconstruct a "whole plant." For museum-based paleobotanists like Seward, the best hope was to find the lucky specimen where the seed was still attached to a leafy branch. As he completed his survey of Yorkshire fossils in museums across Europe, Seward realized more fieldwork was needed. He encouraged a young Welshman, Hugh Hamshaw Thomas, to take up the work. Thomas was quiet and athletic, willing to spend weeks walking the beaches. However, in the generation since William Williamson, many prime fossil spots had been forgotten. On one search of the Gristhorpe Bed, Williamson's most productive site, Thomas had given up hope and was spending the last daylight collecting algae. The seaweed was firmly attached to the bedrock and, pulling with all his might to dislodge it, Thomas yanked apart the rock to reveal the elusive Jurassic fossils. From this locality, he collected hundreds of new specimens and over time realized something that the museum collections could not reveal: some fossils commonly occurred together. For example, the seed-bearing branch known as *Caytonia* occurred with *Sagenopteris* leaves. Thomas hypothesized that these fossils might be parts of the same plant. To test the idea, he looked carefully at the thin, waxy layer, the cuticle, that covered the surface of the plant. The two fossils had the same characteristic arrangement of cells, which meant they belonged to the same plant. Thomas's approach also revealed which plants grew together in ecological communities. Traditionally, paleobotanists grouped species together only by their biological affinities. Thomas documented which species were preserved together, reuniting species into ancient ecosystems. By the end of the twentieth century, nearly 600 plant localities had been discovered along the Yorkshire coast. From these, more than 300 species have been described. And the fossils remain the most complete example of the forest that extended across the middle and high latitudes of the Northern Hemisphere during the Jurassic.

Clues to Darwin's "Abominable Mystery"

In his 1859 masterwork *On the Origin of Species by Means of Natural Selection*, Charles Darwin proposed that lineages of plants and animals could change through time, a concept that today we call "evolution." To Darwin, long stretches of geological time were essential to these transformations. He understood that over many generations, breeders could change the features of animals by choosing which individuals produced offspring. Nature, Darwin proposed, could act like the breeder, influencing, although not with consciousness or intent, which individuals contributed more offspring to the next generation. The traits those successful parents possessed would be passed down, changing species over many generations. Darwin acknowledged that changes from generation to generation were minuscule, and uncountable generations would be needed to accumulate the many differences between, for example, fish, reptiles, birds, and mammals. Darwin knew that the Earth was truly ancient, and there was more than enough time for this "natural selection" to gradually produce all of life's diversity.

Fossils provide evidence of change through time, offering examples of ancient creatures that showed a combination of traits found today in disparate groups. The early bird *Archaeopteryx* found in the Jurassic-age Solnhofen Limestone of Germany supplied a high-profile example. Described just two years after the publication of *Origin of Species*, *Archaeopteryx* combined the features of reptiles (teeth and a bony tail) and of birds (feathered wings and a beak) into an animal that showed the transition between these two groups. Flowering plants presented a problem. The Jurassic fossil record yielded no intermediaries between plants that produced cones, like conifers or cycads, and angiosperms. And when flowering plants first appeared in the Early Cretaceous fossil record, they were already diverse. Either flowering plants evolved quickly, or angiosperm ancestors hid from the fossil record for long stretches of geological time. Darwin disliked both possibilities and, in an 1879 letter to his friend and colleague Joseph Hooker, he wrote: "The rapid development ... of all the [angiosperms] within recent geological times is an abominable mystery."

The fossil record of flowering plants presented two challenges to Darwin's notion of gradual evolution. First, there were few examples of intermediate forms, bridges from the other seed plants to flowering plants. In 1925, Hugh Hamshaw Thomas proposed an intermediary from among the Jurassic fossils of Yorkshire. Thomas had dissected the seeds of *Caytonia*, which seemed to have its seed-bearing structure enclosed in two layers of tissue, a feature unique to flowering plants. Although Thomas knew that *Caytonia*'s leaves and pollen

organs were not angiosperm-like, he speculated that *Caytonia* might be a transitional fossil that combined features from groups that are distinct today. Later, Tom Harris recognized that the *Caytonia* plant produced conifer-like pollen, which took *Caytonia* off the main line to angiosperms. But the notion that there were plants in the Jurassic landscape that might link angiosperms and the other seed plants sparked a new approach.

In the 1980s, researchers applied the new method of cladistics to the problem of angiosperm origins. Cladistics relies on three principles derived from Darwin's theory of natural selection. First, any two groups are related by descent from a common ancestor at some time in the past. Second, new species form from that common ancestor by splitting into groups that become more different over time. And third, new features arise when new species form. Using desktop computers that became available in the 1980s, scientists wrote computer programs that applied these rules to the collection of features possessed by living and fossil plants. Their computers produced patterns of relationships among organisms in the form of branching diagrams called cladograms. Paleobotanists recognized that plants like *Gnetum*, along with the extinct Bennettitales and a Southern Hemisphere plant called *Pentoxylon*, appeared to share a common ancestor with flowering plants. They called this group the "anthophytes." *Gnetum*, for example, has angiosperm-like net-veined leaves and a pattern of fertilization superficially like that of flowering plants. The Bennettitales, which were diverse and well known from the Jurassic, had flowerlike structures with pollen and ovules together and surrounded by petallike modified leaves. However, by the turn of the twenty-first century, new data from detailed analysis of plant DNA revealed that *Gnetum* shared a common ancestor with conifers, not flowering plants. Instead, it appeared, angiosperms and plants like *Gnetum*, the Bennettitales, and *Pentoxylon* were evolving similar traits independently in response to similar evolutionary pressures in their environment.

The flowering plants' second challenge to Darwin's notion of slow and gradual evolution was their apparently explosive diversification in the Early Cretaceous. In the 1970s, detailed study of angiosperm leaf fossils from the Potomac Group rocks of Maryland and Virginia showed that the oldest layers (Aptian age—121–113 million years ago) preserved only a few, relatively simple, leaves and one or two kinds of angiosperm pollen. However, younger layers (Albian and Cenomanian age—100–94 million years ago) contained more kinds of leaves and pollen with more complex forms. Was this the beginning of flowering plant diversity? Probably not. In Albian-age clays from Portugal, perhaps 200 species of tiny flowers were found preserved as charcoal. By the end of the Early Cretaceous, flowers were already diverse.

Meanwhile, botanists studying the rates of evolutionary change in the DNA of living plants concluded that the common ancestor of all living angiosperms likely grew in the Late Triassic or Early Jurassic. The living species most like that ancestor is *Amborella trichopoda*, a rare understory shrub from the highland forests of New Caledonia in the southwestern Pacific. Further analysis revealed that flowering plants evolved slowly at first but then diversified explosively early in the Cretaceous, before the fossil record captured them. Although it appears that Darwin was wrong in insisting that evolutionary change always happened slowly, the origin of the flowering plants remains a mystery. And that is one of the most enticing aspects of paleobotany: Someone, somewhere, could at this moment be splitting open layers of rock to reveal a fossil that will upend everything we think we know.

WHAT PLANTS LIVED IN THE MESOZOIC?

Nearly all Mesozoic plants evolved strategies for thriving in a world full of disturbance. To flourish in the Mesozoic, a plant's ancestors had survived the Great Dying—the mass extinction at the end of the Permian period. This primed Mesozoic survivors to cope with climate extremes. Wildfire was common in the Mesozoic, and plants needed ways to either survive fire or regrow quickly afterward. Gigantic herbivores crashed through the landscape, trampling plants underfoot, and munching vast quantities of foliage. Dinosaur disturbance challenged slow-growing conifers and cycads to defend themselves and pushed other groups like ferns and flowering plants to grow and reproduce quickly.

Ferns and Other Plants that Reproduced with Spores

Among the most diverse Mesozoic plant groups were the "true ferns." Ferns first appeared in the Paleozoic. Their large, complex leaves had simple veins, and their stems consisted of many strands of water-conducting tubes that divided and rejoined in complex patterns. This flexible design allowed ferns to evolve a wide range of lifestyles. Some fern stems grew upright in tree form; others crept along the forest floor, producing carpets of interconnected leaves. Some ferns began their lives on the forest floor and climbed to reach the sunlight. Others lived their entire lives on the branches of canopy trees. Like most Paleozoic plants, ferns of the Mesozoic reproduced with spores. These germinated into tiny, tongue-shaped plants called gametophytes. The fern gametophyte lacked leaves, stems, or roots and did not have specialized water-conducting cells. Instead, it produced egg cells and sperm—gametes. Sperm swam through the environment in a film of water to reach the egg cell, and after fertilization, the embryo grew into a new leafy fern.

In addition to the true ferns, several other spore-producing groups inhabited the wet forests of the Mesozoic. Mosses, known from tiny leaves and spore-producing structures trapped in amber, clung to trees in the wettest and shadiest spots of the forest much as they do today. Lycophytes, descendants of the great trees of the Late Paleozoic coal swamps, were rare. Most tree-sized lycophytes became extinct before the Mesozoic, but their smaller cousins persisted in wetlands worldwide. Horsetails also populated streamsides and wetlands, occupying the environments in which they still grow today.

Cycads

Cycads arose in the Early Permian and survived the end-Permian extinction to become one of the most abundant plant groups of both the Triassic and Jurassic. Woody leaf bases protected the cycad stem from fire and hungry predators. The trunks could be short and rounded or tall and branching, but few cycads grew to tree-size. Cycad leaves were heavily armored with tough outer layers, a formidable array of protective chemicals, and mouth-piercing spikes. Because individual leaves could live for several years, they needed to be well defended against herbivores and disease. Unlike ferns, cycads reproduced with seeds. Seed plants encase egg cells in parental tissue to produce a new structure called an ovule. The cycad ovule was displayed on a stubby leaf where it awaited pollen produced by other cone-like structures. Chromosomes received from a cycad's parents determined the plant's sex, so an individual cycad produced either seeds or pollen, but not both. Pollen must reach the ovule to fertilize the egg cell within to make an embryo. For many years botanists assumed that cycads relied on the wind to transport their pollen. Closer examination revealed that insects like beetles and flies distributed most cycad pollen. To entice their animal partners, pollen-producing cycads provided food and shelter. Seed-bearing individuals made a sweet liquid at just the spot where pollen must be deposited. Once mature, cycad seeds also used animals to get around. Bright red, yellow, or orange seeds attracted dinosaurs—as they do birds today—who swallowed the seeds and later deposited them in their dung.

Conifers

Like many Mesozoic groups, conifers first appeared in the Paleozoic, but most of the groups familiar today arose in a series of species-producing pulses in the Triassic and Jurassic. Conifers found ecological success by elaborating on a simple design. Most were trees with tall, straight trunks and branches that emerged at right angles. Today, the world's largest tree (*Sequoiadendron giganteum*, the giant redwood) and tallest tree (*Sequoia sempervirens*, the coast redwood) are both conifers. Most Mesozoic conifers had straight trunks with relatively simple wood infused with sticky, aromatic resin, which both rendered them unpalatable to herbivores and helped seal and disinfect wounds when they occurred. Some conifers also covered their stems with thick, fibrous bark that insulated growing cells from heat, fire, and insect attack. Most conifers had small, tough, waxy leaves that persisted from year to year. Because leaves were long-lived, they had physical and chemical defenses to deter herbivores. Nonetheless, long-necked sauropod dinosaurs like *Camarasaurus*

and *Brachiosaurus* stripped leaves from conifer canopies with their rake-like front teeth. Like cycads, conifers reproduced with seeds. Ovules and pollen developed in separate cones; pollen cones developed at the top of the tree where the copious pollen they produced could catch the wind, and woody seed cones formed on lower branches where they would hold seeds until they were ready to disperse. Some conifers sealed their seed cones with resin. These cones only released seeds when wildfire melted the seal, and the passing blaze primed the soil for germination.

Ginkgos

A diverse collection of ginkgo species populated Mesozoic forests. Although they were never abundant, ginkgos lurked in many Mesozoic floras. Most ancient ginkgos resembled the modern *Ginkgo biloba* with straight, conifer-like trunks, simple wood, and fan-shaped leaves. Mesozoic ginkgos reproduced with seeds. Like cycads, the modern *Ginkgo* has separate pollen- and seed-producing individuals; the fossil record has not yet revealed whether this was true of ancient ginkgos too, but their shared ancestry with cycads suggests that they did. Pollen-producing ginkgos made delicate clusters of pollen sacs that dangled in the wind. Pollen blew to seed-producing trees that displayed their ovules on short stalks. After pollination, the outer part of the ginkgo ovule swelled into a fleshy, fruit-like cover, surrounding a hard shell that protected the developing embryo within. The large size (about 2 cm/0.75 in) and complex structure of ginkgo seeds suggest that they may have been swallowed and dispersed by animals. However, no ginkgo seeds have yet been found in the stomach or dung of any Mesozoic animal.

"Seed Ferns" and Other Enigmatic Plants

In addition to plants familiar to today's gardener—like ferns, cycads, conifers, and ginkgos—an assortment of other seed plants flourished in Mesozoic forests. Many flirted with the features that would later come together in flowering plants. For instance, so-called "seed ferns" produced fernlike leaves with seeds instead of spores.

Another group, the Bennettitales, masqueraded as cycads, with stout, sparsely branched stems armored with woody leaf bases. Their leaves also superficially resembled those of cycads, although details revealed that they were something else entirely. Bennettitales produced flowerlike reproductive structures that combined pollen organs and ovules, surrounded by petal-like leaves. Bennettitales "flowers" also functioned like true flowers, attracting insects that snacked on pollen before moving on and distributing pollen among plants. Yet another group, the gnetophytes, produced net-veined leaves and specialized water-conducting cells in their wood, much like flowering plants. While gnetophyte reproductive structures looked more like those of conifers or ginkgos, the details of fertilization and embryo development paralleled that of flowering plants in some ways. For a while, these similarities led paleobotanists to suggest that gnetophytes, which appeared in the Triassic, might have shared a Permian or Triassic common ancestor with flowering plants. However, genetic studies revealed that the gnetophytes were most closely related to conifers. This left an inescapable conclusion: The Mesozoic environment pushed many plant lineages toward the suite of characters that, when they finally all came together in flowering plants, catapulted this lineage to world dominance.

Flowering Plants

Amid the lush and diverse Jurassic forests, a botanist might not notice the small, woody shrubs growing on mounds of exposed earth churned up by a fallen tree or passing dinosaur. These plants combined features that allowed them to exploit disturbed habitats unavailable to slow-growing Mesozoic seed plants. Flowering plants had specialized water-conducting cells and net-veined leaves that helped them move water quickly from root to leaf. More efficient water transport boosted the rate at which plants could turn carbon dioxide into sugar through photosynthesis, allowing the early angiosperms to take advantage of every fleck of sun that trickled through the conifer canopy. These newcomers did not invest in armored stems or reinforced leaves. Instead, they built their leaves cheaply and replaced them quickly if they were damaged or consumed. Completely

enclosing their ovules in a second layer of the parent's tissue protected delicate egg cells and allowed the parent to control which grain of pollen fertilized the egg. This opened a new range of ways to generate genetic diversity. Flowering plants also took their ancestors' relationships with pollinators and seed dispersers to a new level, developing complex and specialized partnerships that helped define species and allowed mating between individuals in widely separated forest patches, both of which spurred diversity. But for a long time—perhaps tens of millions of years—this new group hid somewhere out of sight of the fossil record. Then, in the Early Cretaceous, a few angiosperm species developed the capacity to photosynthesize in full sun and colonized frequently flooded riverbanks. More species followed. Diversity increased rapidly, but at first, flowers remained uncommon. Slowly, angiosperms expanded out of the tropics, following warm climates poleward during geologically brief bursts of hothouse climate. This vanguard colonized ephemeral spaces like lakeshores and riverbanks, later spreading into established plant communities where they crowded out their neighbors. Flowering plants could grow more, reproduce faster, and seemingly disperse farther than their seed plant cousins. And by the end of the Mesozoic, they had transformed the world.

AGE DATING MESOZOIC PLANTS

The Mesozoic era spans 186 million years, from the beginning of the Triassic about 252 million years ago until the end of the Cretaceous about 66 million years ago. The beginning and end of the era were punctuated by two of the five largest mass extinctions observed in the Phanerozoic eon—the last 539 million years of Earth's history. A third mass extinction marked the boundary between the Triassic and Jurassic periods, about 201 million years ago. Together, the Triassic (spanning about 51 million years), the Jurassic (spanning about 56 million years), and the Cretaceous (spanning about 79 million years) make up the Mesozoic. Each of the three periods is divided into two or three epochs, and each epoch is divided into ages, the shortest subdivision of geological time. The periods, epochs, and ages of the

Mesozoic were defined long before scientists knew the true age of the Earth or could measure their spans in millions of years. Instead, time boundaries were defined by events in this history of life—extinctions or first appearances of specific fossils. Because evolution and extinction do not follow predictable patterns in time, the Mesozoic ages may be as short as a million years or as long as about 20 million years and no two have the same length. Until the mid-part of the twentieth century, geologists did not know when, in millions of years before today, these events took place. With the advent of numerical age dating using the decay of radioactive elements like uranium and potassium, scientists could tag the fossil-defined geological timescale with absolute age dates, a process that continues to be refined today.

Most plant fossils are preserved in sedimentary rock. Sedimentary rock forms from any other type of rock that is broken apart by physical processes (e.g., waves, wind, and ice) or biological processes (e.g., plant roots). The resulting particles, called sediment, can then move around in the environment. Water is the most common agent for moving sediment—pebbles, sand, silt, or clay—although wind can carry the smallest particles, and glacial ice can carry even house-sized boulders. When sediment finally comes to rest, it can entrap plant parts—leaves, seeds, wood, cones, and even pollen and spores—preserving them as fossils. Sediment may be buried by more sediment, compacted, glued together by minerals in groundwater, and preserved as sedimentary rock potentially for hundreds of millions or billions of years. When paleobotanists find a fossil preserved in sedimentary rock, they use geological clues to age date the rock and the fossil encased within.

Fossils are a first line of evidence in age dating sedimentary rock. Because new species of plants and animals are always arising through evolution and other species are becoming extinct, the combination of species present changes throughout geological time. To be a good time-telling fossil, the organism should be widespread, abundant, and easy to recognize. In terrestrial environments, pollen and spores meet all of these requirements. A teaspoon of sedimentary rock might contain hundreds of thousands of individual pollen or spore grains, making these the most abundant type of plant fossil. Many were dispersed by wind and could be carried for long distances, and others were produced by plants that grew over a wide geographic range and in many types of environments. Such species make excellent time-telling fossils. Pollen and spores are covered with a durable material called sporopollenin that stands up well over time, meaning that pollen and spores are easy to preserve, another feature of a good time-telling fossil.

Unlike igneous rocks—rocks that solidify directly from lava or fall as volcanic ash—sedimentary rocks cannot be dated directly. Geologists use several indirect methods together to date sedimentary rocks and their fossils. Consider an example from western North America where age-dating methods are applied together to refine the age dates of three small quarries yielding fossil leaves. In one quarry, sediments preserve the pollen species *Proteacidites retusus* from the Late Cretaceous Coniacian to Maastrichtian ages, and in another quarry, *Aquilapollenites quadrilobus* from the Late Cretaceous Campanian to Maastrichtian ages has been found. Neither species occurs in rocks from the Cenozoic. When found together, these two species indicate that a sedimentary rock and any fossils it contains can be confidently dated to the latest Cretaceous, sometime between about 84 and 66 million years before the present. In contrast, the third quarry yields *Brevicolporites colpella* pollen that is found only in Cenozoic rocks. From this evidence a paleobotanist can conclude that the first and second quarries are of Cretaceous age and the third quarry is Paleogene.

However, fossil age dates require careful scrutiny. For example, a key species can provide a good age estimate, but absence of that species does not always mean the rock is not of the indicated age. Characteristic pollen species might be missing because their parent plants grew elsewhere or the pollen was not preserved. In addition, durable pollen grains erode from older sedimentary rock and may be transported and redeposited with younger sediments, making the sediment appear older than it really is. Rivers flowing over sedimentary rocks of Cretaceous age may erode and mix Cretaceous pollen grains with younger pollen species.

Once fossils provide a ballpark age, tools like magnetic age dating can refine the estimate. Changes in Earth's molten core occasionally reverse the polarity of the planetary magnetic field. Microscopic mineral grains in rocks act like tiny compass needles, aligning with the magnetic field at the time they were formed. Magnetic field reversals happen in a geological instant and at irregular intervals and affect rocks all over the world at the same time, making them good global time markers. However, a period of normal or reversed magnetic polarity will appear the same in the Triassic as it does in the Cretaceous, so magnetic age dating works best when fossils have already narrowed the possibilities. In the western North America quarries example, several magnetic reversals happened during the time when *Aquilapollenites quadrilobus* lived. However, only one reversal straddled the transition between Cretaceous-age *Aquilapollenites*

quadrilobus and Paleogene-age *Brevicolporites colpella*. At our fossil site, magnetic sediment grains in the bottom and top of the outcrop show normal polarity. However, a section in the middle indicates reversed polarity. When combined with the age estimates derived from pollen, geologists conclude that the reversed polarity interval represents the very latest Cretaceous and earliest Cenozoic.

At the most detailed time resolution, the order of sedimentary layers can reveal age. Sedimentary layers accumulate with the oldest layers on the bottom. Therefore, the three fossil quarries can be arranged in time from oldest (Quarry A) on the bottom to youngest (Quarry C) on the top. When coupled with the magnetic and

Fossil leaves have been found in three quarries at this site. The fossils can be age dated by combining clues from pollen preserved in the sediment, radiometric dates from volcanic ash, and changes in Earth's magnetic field, noted by black (reversed polarity) and white (normal polarity) bars.

pollen data, the two lowermost quarries are latest Cretaceous in age and the uppermost is Paleogene.

In very special circumstances, radioisotopic age dating can help date fossils. This method relies on the decay of radioactive elements like potassium and uranium trapped in mineral crystals. Radioisotopic dating—sometimes called radiometric dating—works when liquid rock cooled and trapped radioactive atoms in mineral crystals. In cases where a layer of volcanic ash was sandwiched between layers of sedimentary rock, precise age dates from the volcanic material can bracket the age of a fossil. In the western North America example, radioisotopic dates in two volcanic ash layers provided ages of 66.04 million years and 65.99 million years bracketing Quarry C. Although the magnetic reversal was preserved in sedimentary rocks at our fossil site, at other places around the world, it occurred in igneous rocks that could be dated using radioactive isotopes. Since magnetic reversals happen at the same time planet-wide, we can use those dates—66.31 million years for the beginning of the reversal and 65.72 million years for the end—to further refine the age of our fossil sites. Combining all this evidence, a paleobotanist can conclude that fossil leaves preserved in Quarry A are older than 66.31 million years and those in Quarry B lived between 66.31 and 66.04 million years ago; together, they provide a glimpse of the vegetation from the last 250,000 years of the Mesozoic. The fossil leaves preserved in Quarry C lived between 66.04 and 65.99 million years ago, or during the first 50,000 years or so of the Cenozoic.

GEOGRAPHY AND CLIMATE

As it does today, climate in the Mesozoic varied with latitude and geography. Areas closer to the equator tended to be warmer and less seasonal than those closer to the poles. Similarly, the interiors of continents and the downwind side of mountain ranges tended to be drier than coastal areas. High mountains and expansive continents created monsoon circulation in which large, wet, low-pressure systems drenched some areas with seasonal rain. This regional variability played out in a Mesozoic atmosphere that was enriched in the greenhouse gas carbon dioxide and therefore was generally warmer than today's.

At the dawn of the Triassic, carbon dioxide levels may have topped 1,000 parts per million by volume (ppmv). For comparison, preindustrial carbon dioxide levels in the modern era were about 280 ppmv and topped 400 ppmv in 2012. Greenhouse gases at Triassic levels produced an ice-free warmhouse climate worldwide. While the poles might have experienced freezing winters, in a warmhouse world, snow could not have persisted all year even in the highest mountains. By the Middle Triassic, the global climate had cooled such that the poles experienced a winter freeze. Temperatures warmed through the Early Jurassic then tumbled into a brief Icehouse in the Middle Jurassic. Climate warmed in the Late Jurassic when carbon dioxide levels were again around 1,000 ppmv and remained there until the beginning of the Late Cretaceous, when levels found a new steady state around 500 ppmv and Earth returned to the warmhouse in which polar winters were chilly but glaciers were absent.

Geography enhanced global climate trends. At the beginning of the Triassic, most of Earth's landmasses remained connected as the supercontinent Pangaea. Pangaea extended from pole to pole and was perhaps 9,000 km (almost 6,000 mi) from east to west at the equator. Great equatorial deserts formed in the center of the enormous continent, with a warm temperate climate and extensive forests in the middle and high latitudes. In coastal zones, powerful monsoons brought torrential rains. The pattern intensified as carbon dioxide levels rose through the remainder of the Triassic, with great floods periodically washing forest trees downriver. As the Triassic ended, forces deep in Earth's core set off a geological chain reaction that broke the supercontinent into pieces. What would become the Atlantic Ocean opened as North America and Africa began to unzip from north to south. The split created rift valley lakes along the eastern coast of North America, and proximity to the opening ocean brought seasonally wet climates to the region. Throughout the Jurassic, the interiors of continents remained dry, and the middle to high latitudes in both hemispheres stayed temperate and damp.

By the Early Cretaceous, Earth's landmasses had separated into the two great continents that dominated the rest of the Mesozoic: Gondwana and Laurasia. Gondwana, a supercontinent in its own right, included present-day South America, Africa, Antarctica, India,

Early Triassic

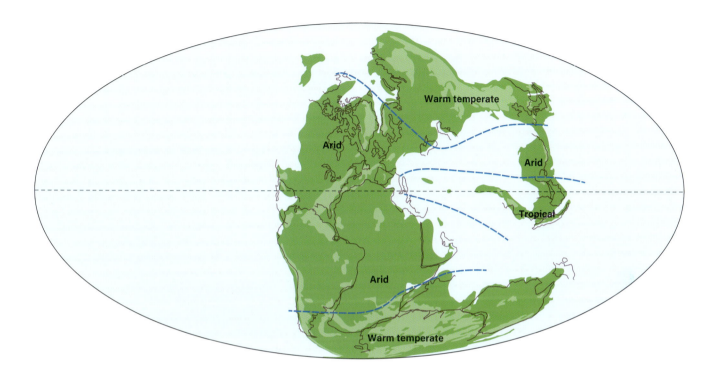

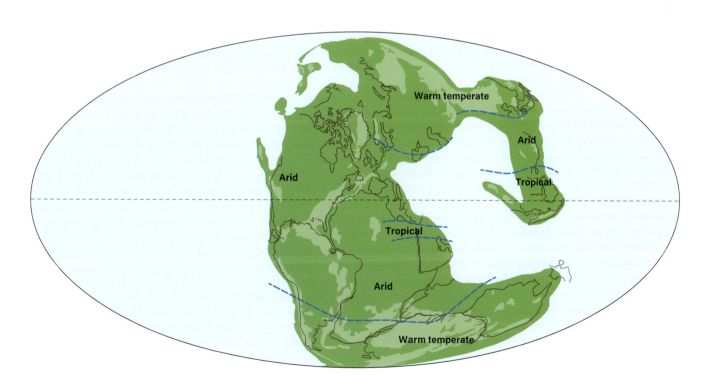

Middle Triassic

Late Triassic

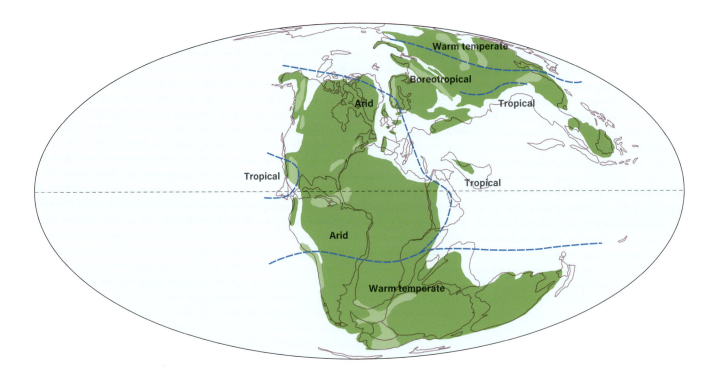

Early Jurassic

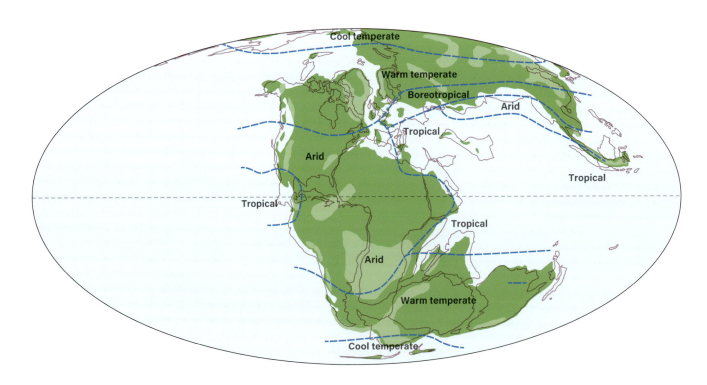

Middle Jurassic

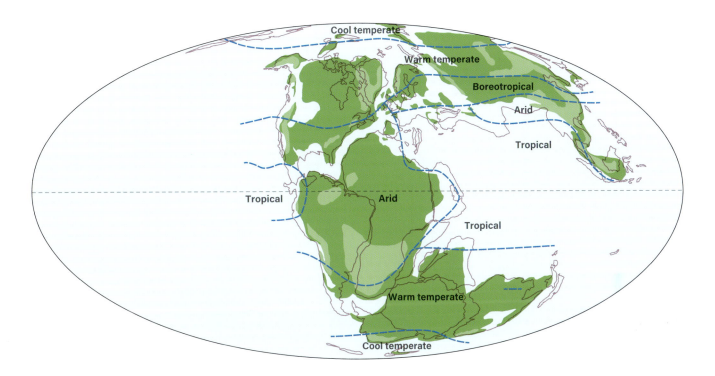

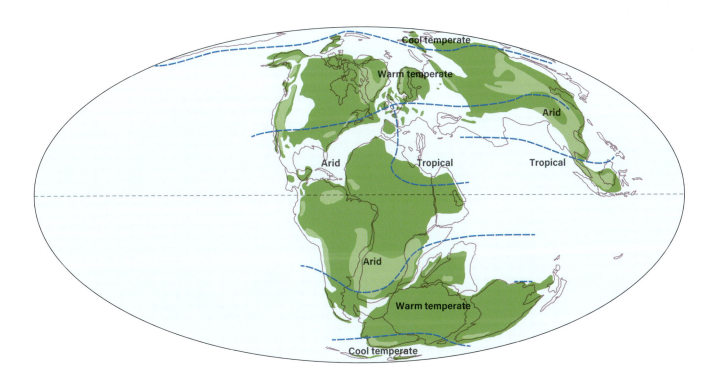

Late Jurassic

Early Cretaceous (Berriasian–Barremian)

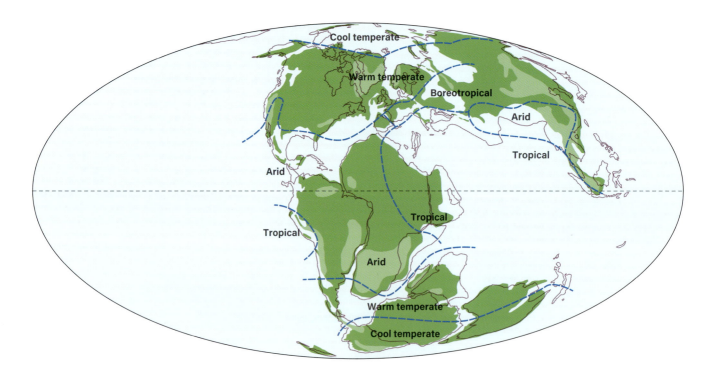

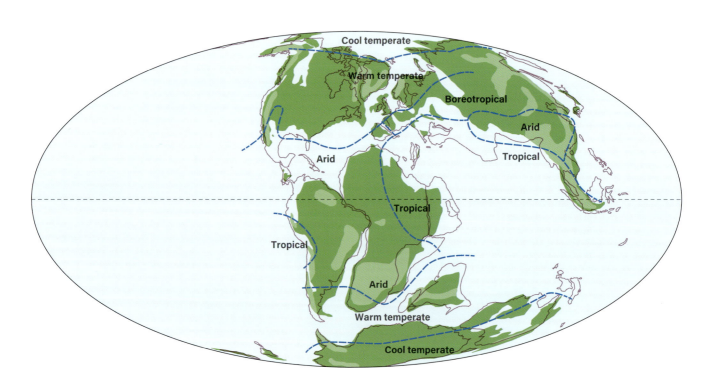

Early Cretaceous (Aptian–Albian)

Late Cretaceous (Cenomanian–Coniacian)

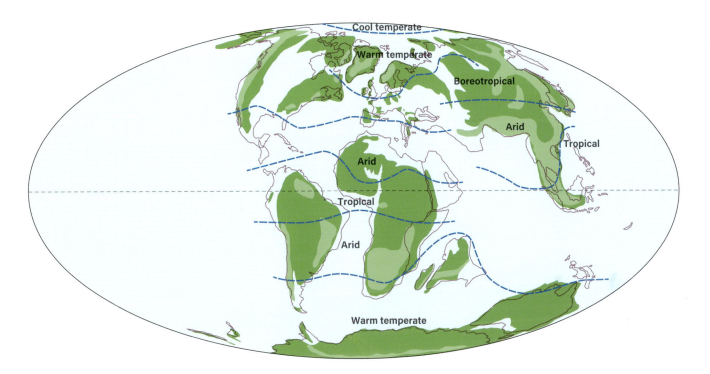

Late Cretaceous (Santonian–Maastrichtian)

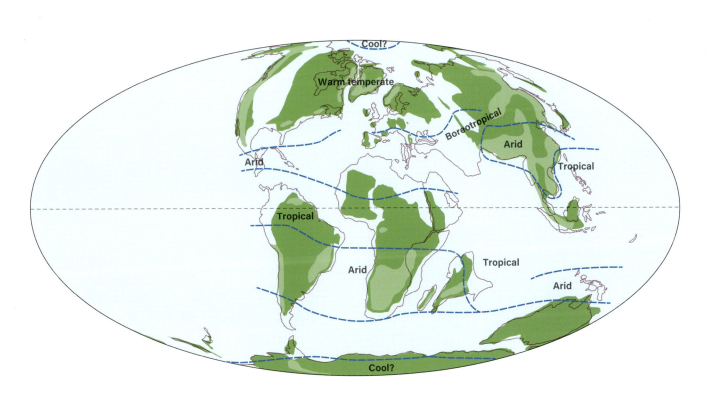

New Zealand, Madagascar, and Australia and remained isolated until the end of the Mesozoic. Laurasia's eastern shores received abundant moisture from prevailing winds that blew across the Tethys Ocean, allowing wet tropical climates to extend to nearly 60° north and south latitude, while the interior and western sides of the continent stayed dry. Warm temperate forests spread across the middle latitudes, and as carbon dioxide levels dropped, the poles once again experienced a winter that was not only dark but also cold. By the Late Cretaceous, shallow seas pushed into the interior of most continents and delivered regular rain to coastal areas. Latitudinal patterns of precipitation similar to those experienced today began to emerge. Rain fell daily in tropical latitudes, while the subtropics experienced wet and dry seasons. Middle latitudes were warm and moist, and cool temperate conditions prevailed at the poles.

EXTINCTION AND RECOVERY

The Mesozoic was bound by two of the "Big Five" mass extinctions of the Phanerozoic eon and included a third. The first of these extinctions took place 250 million years ago at the end of the Permian period and marked the beginning of the Mesozoic. The end-Permian extinction, which paleontologist Douglas Erwin called the Great Dying, resulted from enormous volcanic eruptions in Siberia that pumped the gases carbon dioxide and sulfur dioxide into the atmosphere as lava spilled out of long fissures in Earth's crust. Estimates suggest that these volcanic vents spewed more than a cubic kilometer of lava each year for more than a million years. The carbon dioxide they belched intensified the Permian greenhouse and raised global temperatures rapidly. Extinction resulted not just from the magnitude of this climate change but from its rapid pace. These volcanic gases also acidified the ocean and changed its chemistry in other ways that were toxic to most creatures. In the ocean, the end-Permian extinction wiped out about 87% of all genera, including, for example, all trilobites, 98% of snails, and 96% of corals. On land, eight of the nine orders of insects became extinct in what remains the largest insect mass extinction the world has seen. Perhaps half to two-thirds of

land vertebrates also became extinct, with the greatest losses among large herbivores.

The plant story is more difficult to discern. The supercontinent Pangaea assembled during the Permian, creating large regions of dry climate and high elevation where plants probably lived but seldom became fossils. The wetlands, streamsides, and floodplains where fossil plants usually preserve dwindled in many areas. Without an extensive fossil record, assessing Late Permian plant diversity, and thus the degree of extinction across the Permian–Triassic boundary, remains a challenge. However, microscopic pollen and spores provided important clues. At the global scale, pollen and spore diversity stayed steady across the Permian–Triassic boundary, with extinctions being nearly matched by first appearances of new species. Of the groups that survived the extinction, ferns and cycads were the only ones to show notable loss of diversity across the Permian–Triassic boundary, but within the first million years of the Mesozoic, ferns and cycads embarked on a period of rapid diversification, replacing and exceeding their pre-extinction diversity.

A different picture emerges when we look specifically at equatorial latitudes. There, the rich conifer forests of the Late Permian disappeared completely and suddenly, followed by a brief interval dominated by a spore called *Reduviasporonites*. This spore was originally thought to belong to fungi that feasted on the dead and dying conifers. However, subsequent analysis suggested that these fossils represented hardy algae living in the stagnant water. Conifer forests were quickly replaced by the lycophyte *Pleuromeia*. For nearly five million years, *Pleuromeia*, a shrub-sized lycophyte with ancestry among the giants of the Paleozoic coal swamps, grew in nearly single-species stands across a swath of the tropical lowlands. However, when climate cooled a little, conifers, cycads, and ginkgos migrated back into equatorial regions and reestablished the forests that thrived through the Triassic. Several groups traded dominance in the Triassic forests for millions of years after the extinction, highlighting a key feature of the way plants respond to catastrophic environmental upheaval: Although many species hung on, plant communities struggled to regain stability for millions of years.

The next mass extinction occurred at the end of the Triassic period, about 201 million years ago. The

Triassic–Jurassic extinction was also triggered by volcanoes, this time linked to the breakup of the supercontinent Pangaea. Rifts that ran from present-day Iceland to central Brazil erupted enormous quantities of lava over about 500,000 years. With the lava came another belch of carbon dioxide that raised global temperature. In the oceans, about 30% of genera became extinct, and coral reefs perished. On land, the last Paleozoic vertebrates died, paving the way for the rapid diversification of dinosaurs in the Jurassic. Mammals and crocodiles suffered little extinction. Globally, plant diversity remained steady, but 15% to 80% of species turned over, suggesting that end-Triassic environmental change reorganized terrestrial plant communities. For example, in the middle latitudes of today's Europe, few species were lost, but the old Triassic ecosystems rearranged as plants followed their preferred climate into new areas. Late Triassic midlatitude forests had a diverse canopy, including the conifer *Podozamites* and the ginkgos *Baiera* and *Ginkgo*. Floodplain wetlands hosted the conifers *Elatocladus*, *Stachyotaxus*, and *Podozamites* and an understory that included several species of Bennettitales and many kinds of ferns. As the Triassic ended, the conifer *Podozamites* accounted for 86% of the trees, and a single surviving bennettitalean genus, *Pterophyllum*, and the fern *Cladophlebis denticulata* dominated the understory. Paleobotanists speculate that rare species and those that required insect pollinators were most vulnerable to extinction across the Triassic–Jurassic boundary. In addition, warmer Jurassic temperatures placed stress on plants with large leaves, including the Bennettitales, and contributed to their decline. In the Early Jurassic, forest ecosystems recovered quickly, with species migrating to follow their preferred climate and evolving quickly to replace lost species.

The final Mesozoic mass extinction brought the era to an end. So many distinctive species were lost that the Cretaceous–Paleogene event defined the end of both the Cretaceous period and the Mesozoic era long before geologists understood the timing and nature of the boundary. Clues to the extinction's cause came first from marine sediments near Gubbio, Italy, where a young geologist named Walter Alvarez found a layer of clay enriched in the rare-earth element iridium. In 1980, Alvarez proposed that the iridium fell with the ashes of a 10 km (over 6 mi) wide asteroid that vaporized upon impact with Earth. Nearly a decade later, geologists revealed a 180 km (over 110 mi) wide crater at the tip of Mexico's Yucatán Peninsula. The asteroid's impact set off a searing atmospheric shock wave and titanic tsunami that spread around the Gulf of Mexico. The shock wave vaporized deposits of the sulfate mineral gypsum in the rocks beneath the impact site. Combined with water in the atmosphere, the resulting sulfate cooled climate and acidified surface water. Compounding the disaster, the impact happened during another great outpouring of lava from volcanic vents located in what is today India. Lava up to 2 km (over 1 mi) thick spread out over about 500,000 square kilometers (193,000 square miles) beginning about 400,000 years before the asteroid impact and continuing for about 400,000 years afterward. Although only about half the size of the Siberian flows that began the Mesozoic, these eruptions boosted atmospheric carbon dioxide and reversed the cooling underway since the mid-Cretaceous.

In the oceans, up to 75% of species became extinct at the end of the Cretaceous, including all of the large marine reptiles like plesiosaurs and mosasaurs, many fish and sharks, and mollusks such as ammonites and belemnites. As many as 98% of reef-forming coral also became extinct. For some marine groups lost in the extinction, such as ammonites and inoceramid bivalves, species began disappearing in the 400,000 years or so preceding the event. For other marine groups, like single-celled plankton, the extinction was sudden and coincided with the asteroid impact. On land, almost all animals larger than about 25 kg (55 lb) became extinct, including dinosaurs, some mammals, and many turtles. However, almost all frogs and salamanders, lizards, and the crocodile-like *Champsosaurus* survived.

Plants recorded the end-Cretaceous impact as an ecological catastrophe. In places as widespread as North America, Antarctica, New Zealand, and Japan, the impact event, recorded by its iridium signature, clearcut and in a few places burned over the latest-Cretaceous forests. Shortly afterward, hardy, disturbance-loving species like ferns and some conifers colonized the denuded landscape and began the process of ecological succession that returned the forests to nearly their former diversity and species composition within a few tens of thousands of years. Clearly, the ecological devastation was patchy because enough individuals survived the catastrophe to reseed the recovering land. Some plants became extinct especially close to the impact site in North America, where up to 57% of species, most of them flowering plants, disappeared. Yet some of these, or at least close relatives that produced familiar pollen types, reemerged Lazarus-like in the Paleocene.

Perhaps the most important Mesozoic plant extinction did not qualify for the mass extinction hall of fame. This extinction happened over a span of about 30 million years as the seed plants of the Triassic and Jurassic—cycads, ginkgos, conifers, seed ferns, Bennettitales, and others—were pushed aside by flowers. Angiosperms probably first appeared in the understory of wet equatorial forests sometime in the Late Triassic or Jurassic, although the exact timing is the topic of active research. By the earliest Cretaceous, flowering plant pollen began to appear in low-latitude sediments. By the Aptian age of the Early Cretaceous, a diversity of flowering plants accounted for about 7% of pollen in tropical latitudes. This nearly doubled by the Albian, as angiosperms spread out of the tropics, marching steadily toward both poles. The flowers churned out new species that boosted diversity significantly. For botanists counting species, the Flowering Plant Revolution of the Late Cretaceous would look more like diversification than an extinction. This was also true for ferns, which diversified beneath the flowering trees, perhaps responding to changes in understory light environments created by dense angiosperm canopies. Meanwhile, conifers, ginkgos, cycads, and Bennettitales—the seed plants of the Triassic and Jurassic—all suffered major extinctions. The slow-growing, slow-reproducing, inefficient seed plants were gradually replaced by flowering plants that could do everything—grow, reproduce, and evolve new species—faster and with the flexibility needed to cope with the rapidly changing Cretaceous world. By the time the end-Cretaceous asteroid hit the Yucatán, the angiosperm conquest of Earth's terrestrial ecosystems was complete. Although a handful of conifers, a few cycads, and a single species of *Ginkgo* hang on into the present, the great conifer forests were gone, until the icehouse came and the highest latitudes grew too cold for most angiosperm trees.

AFTER THE MESOZOIC

In 1912, Walther Gothan coined the terms "Paleophytic," "Mesophytic," and "Cenophytic" as a counterpoint to the animal-centered geological eras: Paleozoic, Mesozoic, and Cenozoic. The Paleophytic began with the evolution of land plants some 450 million years ago and ended during the late Permian, about 260 million years ago. This was the time of spore-producing plants. Seed plants like conifers, cycads, and ginkgos dominated the Mesophytic, which lasted until the middle of the Cretaceous, when the explosive diversification and global expansion of flowering plants ushered in the Cenophytic. Since the dawn of the Cenophytic, flowering plants have been doing what they do best: migrating to track the oscillations of Earth's climate, amassing chemical defenses against predators and pathogens, evolving new ways to grow faster and reproduce more abundantly, and, perhaps most importantly for us, discovering new ways to trick animals into disbursing their pollen and seeds. After the Mesozoic, a tiny, upstart grass that had been sampled by dinosaurs radiated into more than 11,500 species, evolved a whole new strategy for photosynthesis, took over the drying interiors of continents, and set off a cascade of evolutionary change that led to horses, rhinos, bison, cattle, camels, and us. Evolution in the pathogen-filled tropics led other angiosperms to experiment with new compounds that humans discovered could cure disease and dupe the pleasure centers in our brains. And all the while, flowering plants churned out new species at an incredible rate. Meanwhile, a few survivors from the Mesophytic stood quietly in secret valleys and awaited discovery. These are the so-called "living fossils," including *Metasequoia*.

The cypress family (Cupressaceae) included many common conifers of the Mesozoic. Fossils assigned to *Sequoia* were particularly abundant among the dinosaur bones of the Late Cretaceous in western North America. However, some *Sequoia* fossils had arrangements of needles and cones that did not match those of living members. This puzzled paleobotanists, but most accepted that earlier members of a lineage could have combinations of features that did not survive into the modern era. Then, in the early 1940s, Japanese paleobotanist Shigeru Miki discovered extraordinarily well-preserved fossil conifers

The Paleophytic was dominated by a wide range of spore-producing plants, many of which filled the great coal swamps of Europe and North America. These ancient trees stored the sun's energy and the atmosphere's carbon for hundreds of millions of years.

Mesophytic forests (above) were dominated by seed plants like conifers, ginkgos, cycads, and a variety of other extinct groups, with spore-producing plants like ferns in the understory. During the Cenophytic (right), angiosperms took over most terrestrial ecosystems, producing dense, multilayered forests and unprecedented diversity.

in three-million-year-old clay deposits of Japan. Clay has an almost magical ability to preserve fossils. Minute clay grains halt decomposition and preserve even microscopic details. Miki noticed that his material, which he initially assigned to *Sequoia* and *Taxodium*, had features that excluded it from both groups. In 1941, he created a new genus in the cypress family, *Metasequoia*. His discovery was overshadowed by Japan's invasion of Guam in December 1941, and few outside of the small Japanese paleobotanical community learned of the new fossil. Two years later, Chan Wang from the Chinese Bureau of Forest Research learned of a strange conifer growing near Moudao in what is today Hubei Province. Wang visited the site and collected a few specimens that he identified as *Glyptostrobus pensilis*, a common wetland cypress. In 1945, Wang gave a few specimens to Wan-Chun Cheng, a professor at the National Central University at Chongqing. Cheng knew *Glyptostrobus* well and realized immediately that this was something new. Cheng, confident that it was indeed a new type of conifer, sent the material to Hsen-Hsu Hu at the Fan Memorial Institute of Biology in Beijing. In late April of 1946, Hu made the connection between Miki's fossil conifer, *Metasequoia*, and the living trees discovered in Moudao. The fossil and living plant were identical in every detail. In 1946, Hu and Cheng collaborated to name the new tree *Metasequoia glyptostroboides*. Hu immediately sent the news to his American colleague Ralph Works Chaney at the University of California, Berkeley. Chaney was an expert in the fossil plants of the Late Cretaceous and Cenozoic and had many connections in China forged during his expedition to Mongolia in 1925 with dinosaur hunter Roy Chapman Andrews. Chaney helped spread the word about the living fossil to English-speaking readers. In February 1948, Chaney and *San Francisco Chronicle* science reporter Milton Silverman traveled to central China to visit several groves of the newly discovered trees. The weather was cold and wet, making roads impassable in places. Chaney became ill and had to be carried by palanquin for part of the journey. But when they arrived, Chaney realized that many of the enigmatic fossils from North America were actually examples of *Metasequoia*.

Paleobotanists worldwide immediately returned to their collections to discover misidentified *Metasequoia*.

In doing so, they showed that *Metasequoia* first appeared about 70 million years ago and was widespread across the middle and high latitudes of the Northern Hemisphere. It survived the end-Cretaceous extinction and remained common until cooling forced *Metasequoia* out of polar latitudes about 34 million years ago. Drying climate in North America forced the species into a coastal refuge in what is today Oregon and Washington States by about 20 million years ago, while it remained widespread across Japan and central China. By five million years ago, *Metasequoia* was extinct in its former stronghold of North America and its range shrank further as continental glaciers embraced the Northern Hemisphere about two million years ago. The species lived on only in isolated valleys of central China where the ice could not find it. Today, seeds collected by Ralph Chaney in 1948 make *Metasequoia* a novelty in botanical gardens and college campuses—a living plant first described as a fossil—a true living fossil.

The popular street and landscape tree *Ginkgo biloba* is another botanical time traveler from the Mesozoic. The first members of the ginkgo lineage appeared sometime in the Permian. They diversified extensively in the Late Triassic to include species with strap-like and fan-shaped leaves and leaves that resembled triangles of delicate fringe. A preference for moist, lowland habitats meant that distinctive ginkgo foliage found its way into most fossil assemblages, revealing a global distribution throughout the Jurassic and Early Cretaceous. Although seemingly present everywhere, ginkgos never dominated the forests in which they grew. When flowering plants began to invade the lowland landscape in the Early Cretaceous, ginkgos were among the groups that declined, becoming rare in both hemispheres by the end of the Cretaceous. Although ginkgos survived the end-Cretaceous extinction event, they had disappeared in the Southern Hemisphere by about 60 million years ago. From that time, they followed *Metasequoia* in a slow-motion retreat, first abandoning the poles as they cooled and the interiors of continents as they dried. Ginkgos disappeared from North America about 15 million years ago, and the last fossils in Europe date to five million years ago. *Ginkgo* had dwindled to a rare relic in what is today China by the time humans arrived there about 50,000 years ago. However, unlike *Metasequoia*, *Ginkgo* seeds contain an edible portion that became both food and traditional medicine across East Asia. The oldest written record of *Ginkgo* cultivation dates from the eleventh-century Song dynasty in China. By then, *Ginkgo* may have been entirely extinct in the wild but was carefully cultivated by humans for food, medicine, and ritual. In 1712, *Ginkgo* was introduced to the West by Engelbert Kaempfer, who illustrated it in his book *Amoenitatum Exoticarum*. Seedlings returned to Britain in the 1760s grew well, and *Ginkgo* became a prized botanical folly for the aristocrats of Europe. During the industrial revolution, city planners realized that *Ginkgo*'s straight trunk and compact root system made it ideal for crowded city streets. Moreover, the species thrived in a wide range of climates and tolerated polluted city air. Today, humans have helped *Ginkgo biloba*, the sole surviving species of this once diverse Mesozoic lineage, recover much of its former geographic range as a captive species.

The conifer family Araucariaceae, the monkey puzzle trees, offers another Mesozoic ghost-turned-survivor. The family first appeared in the Jurassic and was diverse and widespread throughout the period. Like most conifers, its diversity and abundance declined in the Late Cretaceous, but the family held on in both hemispheres until the end of the Cretaceous. Eliminated from the Northern Hemisphere by the end-Cretaceous event, about 40 species survived into the present day, distributed across temperate and tropical parts of Australia, New Caledonia, New Zealand, Borneo, New Guinea, and a handful of other Pacific islands. In 1994, David Noble, an Australian National Parks and Wildlife Service field officer, rappelled into an unexplored glen of the Blue Mountains in New South Wales. On the rocky slopes of the canyon he found 38 trees of a species he did not recognize. Noble knew it as a member of the monkey puzzle family but it was unlike any species that he had seen before. He collected a few dead twigs and cones from the ground. The tree was indeed new and was named *Wollemia nobilis* in honor of its discoverer. Fifty-four-million-year-old fossil leaves assigned to the genus *Araucarioides* resemble *Wollemia*, but the fossils do not contain enough detail for a definitive connection. *Wollemia*'s strongest link to the Mesozoic may be through pollen. A pollen type called *Dilwynites* resembles *Wollemia* pollen and has been found from the Late Cretaceous of Australia, Antarctica, New Zealand, and Patagonia in South America. *Dilwynites* disappeared from the fossil record about two million years ago at the onset of today's icehouse climate. If the link to *Dilwynites* holds up, *Wollemia* would be another Mesozoic survivor that found a refuge in the Cenophytic. The location of the wild stand of *Wollemia nobilis* remains secret to protect the remaining trees. However, the species grows easily from seeds and cuttings, and botanical gardens all over the world are establishing collections of the species so that plant enthusiasts can glimpse a bit of the Mesozoic world.

BIOLOGY

As animals—specifically land mammals—humans see the world from a mammal point of view. Food is something to find, shelter is a necessity, and social relationships are required, particularly if we want to reproduce. And we count on moving through our environment to meet all these needs. Plants have different needs. Plants make their food, regard being shaded by a neighbor as hardship, and need to reproduce while rooted firmly in place. They move fluids around their bodies without hearts, get rid of wastes without kidneys, sense light without eyes, respond to smells without noses, breathe without lungs, and have bodies built to withstand—not hide from—harsh environmental conditions. Plants are so alien that some people may not realize that they are alive. But they are very much alive: sensing, responding to, influencing, and being influenced by their environments.

MAKING FOOD: PHOTOSYNTHESIS

Almost all the energy that powers life on Earth comes from plants converting sunlight, water, and carbon dioxide into the sugars, proteins, and fats that they, and most other living things, use for food. In the oceans, single-celled algae do most of this work. On land, green plants fill the role, making the molecules of life through photosynthesis. Photosynthesis starts with plants harvesting energy from sunlight. Land plants inherited this magic from their green algal ancestors, so they share a suite of colorful molecules used to capture the sun's energy. One such molecule is chlorophyll, which absorbs light in the red and orange (600–700 nm) and violet (370–390 nm) wavelengths of the visible light spectrum. Another family of molecules, carotenoids, soak up wavelengths in the blue part of the visible spectrum (400–500 nm). The gap between these wavelengths—500–600 nm in the visible light spectrum—corresponds to green, the color that land plant leaves appear to our eyes. When light of just the right wavelength strikes chlorophyll, light energy pushes an electron from the magnesium atom at the heart of the molecule out of its regular position. The molecule now has a little extra energy to do chemical work. Shorter wavelengths (blue and violet) have just the right amount of energy to free a hydrogen atom from water and split it into a free electron and its proton. The hydrogen proton makes the part of the leaf where photosynthesis takes place slightly acidic, which helps capture and store some additional chemical energy in the molecule adenosine triphosphate (ATP). Meanwhile, longer wavelengths (red) help pass the liberated electron to where it is needed to create

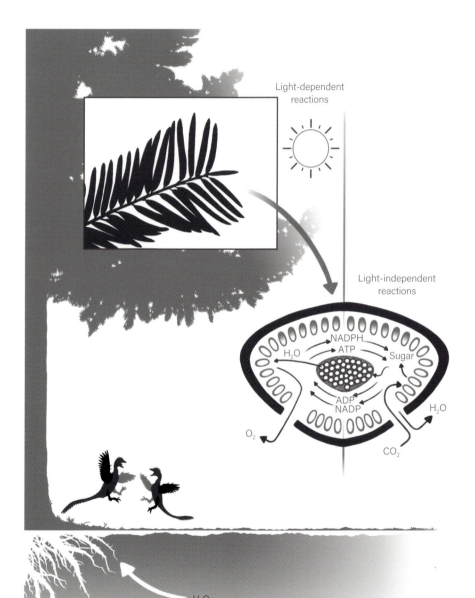

Light-dependent reactions

Light-independent reactions

NADPH
ATP
H_2O
Sugar
ADP
NADP
O_2
H_2O
CO_2

H_2O

In photosynthesis, plants make food using sunlight, carbon dioxide (CO_2), and water (H_2O). The sun splits water to yield an electron and energy. Light-independent reactions transform carbon dioxide into carbohydrates (sugars).

sugar from carbon dioxide. These so-called "light reactions" happen only in sunlight and allow the plant to capture the sun's energy and store it as chemical energy. Freeing hydrogen from water creates oxygen. The plant uses a bit of this oxygen for its own metabolism and releases most of it into the atmosphere. This first step in photosynthesis is responsible for most of the oxygen in Earth's atmosphere. Without this important plant waste product, animal life would be impossible.

The second step in photosynthesis—so-called "light-independent reactions" because they can happen at night—takes carbon dioxide from the atmosphere and converts it to sugar using the chemical energy and extra electron generated during the light reactions. Plants first open small leaf pores called stomata to allow atmospheric gases to flow into the leaf. This refreshes the carbon dioxide near the sites of photosynthesis and releases waste gases like oxygen. Inside the leaf, a protein called ribulose bisphosphate carboxylase-oxygenase ("rubisco" for short) grabs the carbon dioxide molecule and combines it with a molecule of the sugar ribulose bisphosphate and then splits the resulting molecule in two. These molecules pass into a chemical chain reaction where energy (ATP) and the electron from the light reactions help remove one of the oxygen atoms attached to the original carbon dioxide to create a new sugar molecule. Through this process, green plants capture energy from sunlight and store it in the chemical bonds of sugar. The plant will use some of the sugar as its own food, burning some for energy and using the rest as the raw material for other compounds like fats, proteins (including rubisco itself), defensive chemicals, material that makes up the plant's body, and much more. The plant stores any leftovers in its leaves, stems, and roots as starch. Plants also pack starch into seeds to provide nourishment for their embryos as they develop. Humans enjoy some of this starch in the form of wheat, corn, barley, sorghum, millet, oats, and rice.

MOVING FLUIDS WITHOUT A HEART: WATER AND SAP

Some coast redwood trees (*Sequoia sempervirens*) can tower well over 100 m (almost 330 ft) tall. Because water

is essential for photosynthesis, leaves at the tops of trees need a steady supply from the soil in order to harvest energy from light. Leaves in the sunshine also need water to cool themselves—much like animals sweating. In most animals, fluid moves around the body through flexible tubes and is pushed by some sort of muscular pump. Plants have neither flexible tubes—plant cells have stiff cell walls that make them strong and rigid—nor muscular pumps, so evolution found a different mechanism to move water from the soil to the tops of the tallest trees.

The tubes that carry water in most plants are called tracheids. They are single, elongated cells about 10–25 μm (micrometers) in diameter (for comparison, fine-textured human hair is about 20 μm in diameter). Tracheids carry water only after the cells die. They begin like any cell, first dividing and forming a cell membrane that contains all the contents of the cell, then tracheids—and all other plant cells—surround the cell membrane with a stiff cell wall composed of cellulose. To humans, cellulose is a component of dietary fiber and the main ingredient of cotton and linen. Just before

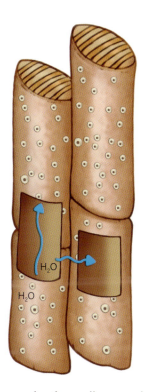

Water-conducting cells, vessel elements (left) and tracheids (right), are closed cells with perforations through which water is drawn from root to leaf.

the new tracheid cell dies, a second wall of a sturdy material called lignin forms outside of the cellulose cell wall. Lignin gives the tracheids extra strength and allows them to support tons of tree. Then the cell dies and the empty tubes are ready to carry water. Ferns, cycads, ginkgos, conifers, and many extinct plant groups have only tracheids to move their water. In contrast, most flowering plants, and a few others like *Gnetum*, evolved wider and longer tubes called vessels. Formed in the same way and also dead at maturity, the average vessel is 40–80 μm in diameter, with the largest 500 μm in diameter and up to 2 cm (over 0.75 in) long. These larger tubes can carry more water faster but, compared to tracheids, more easily lose their water-conducting ability.

The journey of a water molecule from soil to leaf begins with water moving into the roots. Water in the soil moves into the root cell by osmosis in a futile attempt to balance the concentration of sugars and minerals inside and outside root cells. Plant cells rig the game by actively moving minerals like potassium into their root cells. Water naturally tries to balance the concentrations of these minerals, so water moves into the cell to dilute concentrations there. Once inside, the hydrogen atoms in each water molecule form loose bonds with neighboring hydrogen atoms. This connection is weaker than the bonds that hold hydrogen to oxygen in the water molecule, but strong enough to help water move through the plant's system of conducting tubes. Think of water molecules holding hydrogen hands; the bond is sturdy enough to pull water molecules along in a chain, but a strong tug can break the connections. Once inside the root, water molecules join a chain that extends all the way to the very top of the tree.

During photosynthesis, plants open pores on the surface of their leaves to allow carbon dioxide in. At the same time, water molecules leave the leaf. As a water molecule departs the leaf, it pulls the water molecule to which it is loosely bonded and moves the next molecule closer to the pore and the free air outside. This gently tugs the whole chain of water molecules stretching back to the root and draws water slowly up the stem. To help the molecular ascent, hydrogen hands also stick, ever so slightly, to the inside of the tracheids themselves, much like a climber holding the rung of a ladder as they offer a hand to the person below them. In this way, water

leaving the plant through leaf pores draws more water steadily up from the roots. No pump required.

This way of moving water only works as long as water molecules remain connected in a continuous chain. If there is too little water in the roots (drought conditions), too much water leaving the leaves (hot, sunny, or dry conditions), or if ice crystals form in the tracheids themselves (cold conditions), the molecular chain breaks and water can no longer move up the stem. Water-conducting cells in which the chain has broken no longer function for water transport. In most plants, the only solution is to make more tracheids. Woody trees do this throughout their lives. In seasonal environments, they renew their water-conducting cells at the beginning of each growing season. In climates without periodic cold or dry periods, trees continually renew their tracheids and vessels as old cells lose their function. Some plants solve the problem of broken water chains by pushing water up from their roots to refill emptied tracheids and re-form the water chain. When water is abundant in soil, plants capable of this response use energy to concentrate minerals and sugars in their roots and overfill their root cells. These stretched cells generate a positive pressure in nearby tracheids that pushes water up the stem. This so-called "root pressure" cannot lift water all the way to the top of the plant. At most, it can push water a few centimeters to a few meters along the stem. However, root pressure might be enough to recover some tracheids emptied by drought or an unexpected freeze. The larger diameter of flowering plant vessels makes them particularly vulnerable to losing their water-conducting ability in this way. Therefore, root pressure is seen mostly in flowering plants, particularly fast-growers that live in sunny conditions, like tomato (*Solanum lycopersicum*), hops (*Humulus lupulus*), and sunflower (*Helianthus annuus*), and temperate trees whose trunks freeze each winter, such as maple (*Acer*) and birch (*Betula*). Grape (*Vitis labrusca*) vines also experience significant root pressure. Their slender, woody stems have fewer water-conducting cells than a larger tree, so refilling conduits lost to winter's cold is essential to support rapid growth in the spring.

In addition to transporting water from soil to leaves, plants must transfer the sweet products of photosynthesis from leaves to other parts of the plant where they support growth or are stored for later. Sugars generally travel around the plant in living conducting cells called phloem. Phloem cells use energy to move sugar molecules actively into and around their conducting system. Once inside, sugars travel from cell to cell through the phloem by moving from areas of high concentration, near the leaves in summer or roots in spring, to areas of lower concentration.

LIFE CYCLE AND REPRODUCTION

For most animals, reproduction happens when one single cell (sperm) meets another single cell (egg cell) to make a new single cell (zygote) that divides and grows into an individual with trillions of cells containing genes from both parents. Plants certainly can follow the sexual playbook, but many also reproduce without sex by continually growing new shoots from the parent plant. And plants effortlessly switch between these two styles of reproduction depending on the demands of the environment.

To animals used to a single coherent body, plant growth and asexual reproduction can seem confusingly similar because many plants continue to grow throughout their lifetime. A vine, for example, grows from its tip and also makes new branches that can explore the environment in search of light, nutrients, and support. Some plants, including some ferns, horsetails (*Equisetum*), grasses, ginger (*Zingiber officinale*), bamboos, mint (*Mentha*), and aspen (*Populus*) trees, send specialized stems through or along the surface of the soil in search of space to grow, light, water, and mineral nutrients. From the animal perspective, each vertical stem or clump

of leaves appears to be an individual plant, but begin to pull, and you will quickly realize that they are interconnected—all part of a single plant "body." A single quaking aspen (*Populus tremuloides*) may grow 40,000 vertical stems and cover more than a hundred acres. To the animal's eye, these constitute a forest of individual trees. In reality, they are a single, interconnected organism. Other plants, such as jonquils (*Narcissus jonquilla*) and day lilies (*Hemerocallis*), produce underground buds that grow from the parent into a new plant that, if separated, will go on to live independently. And many trees (like *Ginkgo* and redwoods) sprout dozens of new stems when their main stem is damaged. All of these approaches allow plants to reproduce themselves without changing their genetic makeup, an effective approach under favorable conditions. However, growing cells are the most vulnerable to mutation, and growth happens at the tip of each shoot of a branchy plant. Therefore, all of the shoots are similar but not genetically identical. In long-lived trees, this can lead to differences in disease resistance and growth rate among shoots of the same plant.

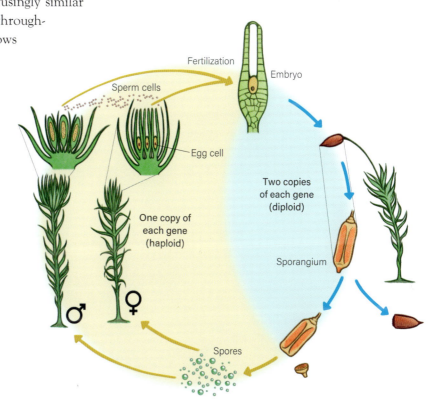

The life cycle of a moss. Portions in the yellow part of the diagram are haploid, containing only one copy of the species' genes. Parts in the blue portion contain two copies of each gene, the diploid.

Shuffling genes through sex produces new combinations of traits, some of which may be beneficial in new or changing environments, so having the option of sexual reproduction has important advantages. New combinations of traits are the raw material of evolutionary change and also help plants cope with environments that vary in space and time. The basics of plant sex work in the familiar way: a sperm cell containing a single copy of each of the parent plant's genes (haploid) combines with an egg cell that also contains a single copy of each gene (haploid). Once fused, the new cell contains two copies of each gene (diploid) and can divide and grow as a new and genetically unique individual. In animals, sperm and eggs are the only haploid part of the life cycle; plants have elaborated considerably on this theme.

The common ancestor of all land plants was a freshwater green alga that flipped the script on sexual reproduction. The green photosynthetic body of these algae contained only one copy of each of the species' genes—they were haploid. When it was time for sex, cells in

the plant body simply transformed into sperm and eggs, which united to form a diploid zygote. The zygote immediately divided without making an additional copy of its genes and these new haploid cells proliferated into new haploid plants. However, because genes from sperm and egg shuffle in the zygote, the offspring had a different combination of genes than either of its parents.

As green algae made their first forays onto land 480 million years ago, they had several difficult problems to solve. First, algal ancestors were constantly bathed in water, but land plants needed a waterproof covering to prevent drying out in the air. Second, pioneering land plants needed to protect their cells from the mutating power of ultraviolet light from the sun. When they lived submerged, water effectively screened out these harmful rays, but on land, cells needed sunscreen. Third, land plants required ways to move water to the sites of photosynthesis. In an aquatic environment, water was easily available everywhere. And fourth, without the buoyancy of water, land plants needed a stiff internal structure to stand up and display photosynthetic surfaces to the light. Solutions to all of these problems involved synthesizing a new array of complex compounds. A large new molecule called cutin and several types of waxes intertwined to provide the waterproof covering, sporopollenin protected sensitive reproductive structures, and lignin reinforced the conducting tubes, making them stiff enough to transport precious water and hold the plant upright. In each case, only diploid cells contained all the instructions needed to synthesize these complicated molecules. The first pioneering land plants probably looked something like moss, in which the green part of the plant remained haploid, like its algal ancestors, with only one copy of each gene. The haploid plant produced sperm and eggs, which, after their union, developed into a small extension of the parent with the biochemical talents that came with being diploid. This allowed the diploid part of the plant to coat its spores in

Egg cell

Spores

Sperm cells

Fertilization

Sporangium

One copy of each gene (haploid)

Embryo

Two copies of each gene (diploid)

The fern life cycle has a free-living haploid (gametophyte, yellow portion of the diagram) and diploid (sporophyte, blue portion). The leafy part that we see is the sporophyte.

sporopollenin and increase the likelihood of survival—and dispersal—in the dry, sunny world. By 412 million years ago, these two so-called "generations" of the plant life cycle—one haploid and one diploid—separated into two independent plants that photosynthesized and lived freely in the environment. The haploid plant produced eggs and sperm that swam through dew to unite and form a zygote that grew into a similar-looking plant (the diploid generation) with two copies of each gene. The reproductive cells in the diploid plant divided without making copies of each gene to produce haploid spores, covered in sporopollenin, thanks to the advanced biochemical abilities of the diploid parent. Spores dispersed widely, protected by their sturdy coating, to grow into a new haploid plant. By Mesozoic time, the biochemical advantages of having two copies of each gene pushed plants to favor the diploid part of the life cycle. Plants like ferns and horsetails spent most of their lives as leafy green organisms in which each cell contained two copies of each gene. When it was time for sexual reproduction, the plant produced spores, each of which contained cells with just one copy of the parent's genes. Spores germinated into small, flat, lobed beings just a few millimeters long that lived almost invisibly in soil and leaf litter. When mature, these tiny plants produced sperm that swam through a thin film of water to unite with the egg produced by a neighbor. The resulting zygote grew into a new, large, leafy diploid plant.

Depending on a film of dew for the union of sperm and egg cell made it challenging for spore-producing plants to practice sex in dry environments. Evolution answered by experimenting with plants that kept the sperm- and egg-producing life cycle stage firmly attached to and protected by the parent plant. These were the first seed plants, lineages that flourished in the drying climates of Pangaea. But the

seed plant solution presented a new problem: matchmaking for sperm to egg. Evolution responded by whittling down the sperm-producing parts to just a few cells, encasing them in a durable sporopollenin shell, and sending them off on the wind—as pollen. With luck, pollen landed on or near the egg-producing structures, and the pollen grain grew a tube to deliver sperm to egg. Further evolutionary refinement coerced animals to act as couriers, increasing the chance that the pollen would arrive at just the right spot for fertilization. While many seed plants still rely on wind to transport their pollen, animal partners opened important new evolutionary doors for seed plants.

Flowering plants took the seed plant playbook to another level. Their pollen remained much like that of their seed plant ancestors but they reduced the sperm-producing generation within to just two cells, allowing pollen grains to be smaller and therefore less energetically expensive to cover in sporopollenin. Likewise, evolution trimmed the egg-producing generation in flowering plants to about seven cells and enclosed them completely in two layers of parental tissue. Besides

Seed plants modify the spore-plant life cycle by eliminating the free-living gametophyte. Seed plants hold egg cells on the parent plant and release sperm within pollen.

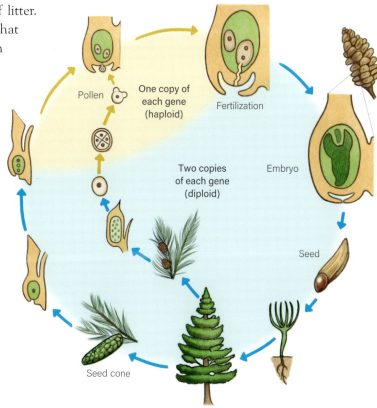

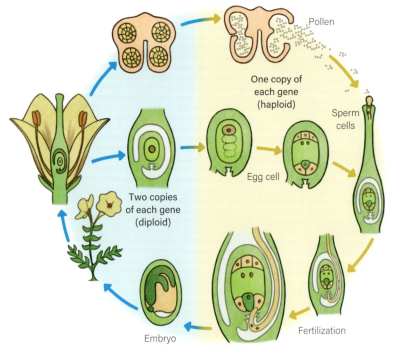

Pollen

One copy of
each gene
(haploid)

Sperm
cells

Egg cell

Two copies
of each gene
(diploid)

Embryo

Fertilization

Flowering plants reduce the egg-producing
stage of the life cycle to just a handful of cells
that are completely protected by the parent
plant. Pollen carries sperm to potential mates.

in the reproductive cells. For animals, this commonly leads to the death of the zygote. However, plants generally embrace the new combination of genes and simply grow into a new plant with a new set of traits. In other cases, sperm and egg cells from different species unite to create a new plant that combines the traits of both. Plant breeders exploit this ability to create new horticultural hybrids with novel features. In the wild, plants that add genes through these reproductive "mistakes" generate more variety upon which natural selection can work. And because the new plant cannot easily reproduce with its parents, a new species is created in an instant. Spore-producing plants regularly use this strategy, with 70% to 90% of species originating in this way. For seed plants the proportion is lower, perhaps 20%, while up to 70% of flowering plant species formed like this.

In addition to creating more individuals, reproduction has the important job of creating new combinations of traits upon which natural selection can work. Evolution allows plants to respond to changing environments and to form new species, thus increasing diversity. Whether through mutations in growing tips, sexual reproduction, or the novel combinations of whole genomes, plants employ a wide range of strategies for auditioning new combinations of traits in an ever-changing world.

SENSING AND RESPONDING TO THE ENVIRONMENT

Viewed through animal eyes, plants are the static backdrop of biology. However, recent research demonstrates that plants sense and respond to their environments in complex and dynamic ways. But plant senses seem entirely alien, and responses happen so slowly that, to animals, plants seem inert. However, plants have

reducing the energy required for sexual reproduction, the angiosperm strategy of surrounding the egg cells completely in two layers of parental tissue meant that the parent plant could evaluate the candidate sperm and potentially stop it from fertilizing the egg. This meant the flower did not waste energy provisioning seeds and fruit that contained embryos that could not survive. This plant version of mate choice also opened up new avenues for natural selection and may have helped rev up diversity during the Cretaceous Flowering Plant Revolution. Yet, the energy angiosperms saved by reducing their egg- and sperm-producing generation, they expended building flowers to attract and reward pollination partners and still more expensive fruit to ensure that their seeds traveled to the best spots for germination. Many credit the advantages that pollination and seed dispersal partners provided for propelling flowering plants to their status as the most diverse and abundant land plants today.

Plants also use some unconventional tactics to add even more variety to their genes during reproduction. Sometimes, the cell division that produces haploid sperm and eggs cells fails, leaving too many copies of each gene

The electromagnetic spectrum, including visible light, the most abundant form of energy emitted by our sun.

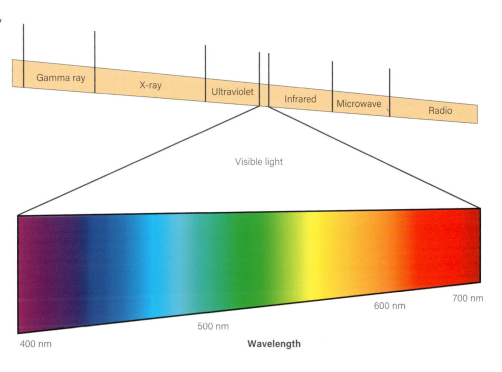

Gamma ray | X-ray | Ultraviolet | Infrared | Microwave | Radio

Visible light

400 nm   500 nm   600 nm   700 nm

Wavelength

sophisticated mechanisms for detecting light, gravity, nutrients, temperature, touch, and a wide range of chemical signals. They use this information, much as animals do, to direct the activities of their lives.

Because light is central to plant life, evolution sculpted a variety of light-sensing mechanisms. The phytochrome system consists of a protein that switches between two forms when exposed to light of specific wavelengths. Phytochrome-r absorbs red light (660 nm) and converts to the active form called phytochrome-fr, which can enter the plant cell nucleus to influence the expression of genes. Phytochrome-fr prompts shoots to grow and leaves to turn toward a light source. Active phytochrome-fr stimulates spore and seed germination so that young plants emerge in favorable light environments. Phytochrome also detects day length and influences seasonal behaviors like flowering at the beginning of the growing season and leaf senescence toward its end. Phytochrome-fr absorbs far-red light (730 nm) and reverts back to the inactive form. Far-red light is most abundant in shady environments, so this system allows the plant to detect when it is under cover. Phytochrome-r also prompts plants to produce leaves tuned to photosynthesis in shade, and in some cases inhibits seed germination. A second group of light-sensing proteins called cryptochromes sense blue (450 nm) and near-ultraviolet (320–400 nm) light. Animals share these photoreceptors, suggesting that they evolved in the common ancestor of plants and animals billions of years ago. Cryptochromes regulate the daily rhythms that set an organism's biological clock. In plants, cryptochromes coordinate with phytochrome to detect day length and coordinate seasonal activities like activating buds in spring, reproduction, and

leaf fall. Cryptochromes also work with phytochrome to help regulate growth; for example, elongating a shoot in shade. By detecting the blue wavelengths of full sun, cryptochromes, along with another group of receptor proteins called phototropins, help plants open stomata to let carbon dioxide in when the photosynthetic machinery ramps up. A final group of light receptors specifically detect ultraviolet light (280–320 nm), which can damage living cells. These receptors trigger DNA repair and the synthesis of sunscreen molecules that protect the leaf from damage.

Plants also need to orient themselves with respect to gravity. Snuggled in dark soil, germinating seeds must sense up, the direction of shoot growth, and down, for roots to penetrate deeper. Plants sense gravity by the position of tiny starch grains in their cells that move around freely in response to gravity. When these granules collide with other cell structures, the plant releases the hormone auxin, which stimulates growth on that side of the cell. As root and shoot emerge from a germinating seed, for example, auxin concentrates shoot growth toward the tip farthest from the starch grains while exerting the opposite effect on the developing root. The gravity-sensing system also allows plants to respond to changes in position by reinforcing support structures and redirecting growth if the plant is tilted.

Although plants make their food—sugars—through photosynthesis, they require a variety of other elements to make the wide range of molecules needed for life. In addition to hydrogen from water and carbon and oxygen from carbon dioxide, plants need nitrogen for proteins, phosphorus for energy storage molecules, and potassium to control water movement in and out of cells. Plants also need calcium, magnesium (for chlorophyll), sulfur, boron, copper, iron, chloride, manganese, molybdenum, and zinc. Roots hunt through soil for these mineral nutrients, detecting them with specialized proteins on their tips. When detector proteins sense needed elements, plant hormones signal a proliferation of root hairs and turn on the absorption system. This moves mineral nutrients into the root, where they ascend the stem to nourish plant cells.

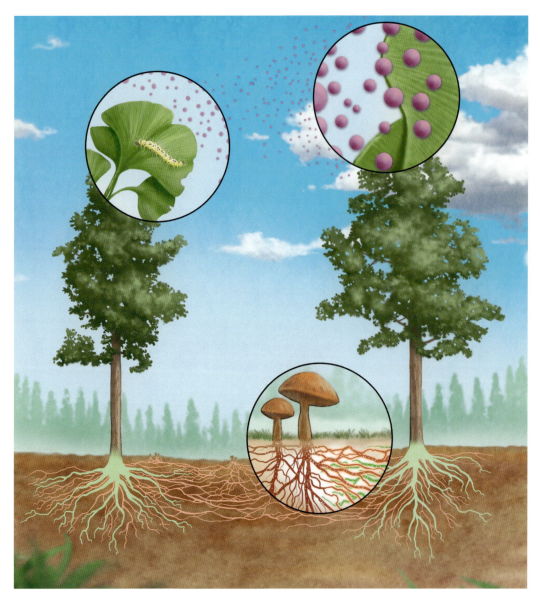

Plants produce and detect chemical signals—scents represented as purple molecules above—that are exchanged among plants, and between plants and fungal and animal partners. Mycorrhizal fungi (drawn in red) facilitate communication among root systems (in green).

Plants sense temperature with other proteins embedded in their cell membranes. These proteins change shape as temperature rises and falls, which stimulates a variety of responses in the plant cell. These proteins work together with light sensors to tell leaves when sunlight causes leaves to overheat and they must open their stomata to cool off. Temperature sensors in other parts of the plant help synchronize growth and other important activities with seasonal conditions.

Plants also sense touch. In animals, detecting pressure relies on specific nerve cells. Lacking nerves, plants perceive touch using springy proteins embedded in their cell membranes. When touched, these proteins change shape and signal physical contact. This mechanism is most important for plants that respond quickly to touch, such as the Venus flytrap (*Dionaea muscipula*), which snaps shut on an unsuspecting insect within seconds of contact. In this case, the plant responds to the fly's touch by sending waves of electrically charged calcium ions through the cell. This signals the plant to move water into cells on the outside of the trap, swelling them and folding the trap closed. In other cases, physical contact changes pressure inside the cell as a whole, triggering slower responses. In climbing plants, for example, a tendril feels the pressure of contact with a nearby surface. The touch causes the tendril to grow toward the pressure, eventually grasping a support.

Plants use other specialized proteins on their cell surfaces to detect specific chemicals. In animals, we call this smell. Like animals, plants sense and respond to compounds that travel through the air. For example, plants emit particular compounds when experiencing drought stress or insect attack. Nearby plants, or different parts of a large tree, detect these chemicals and respond by producing defensive compounds or adjusting water use. In grass, the familiar "new-mown lawn" smell is an alert signal to neighbors that leaves are under widespread assault. Roots also release chemical signals that invite connections with neighboring plants, direct the behavior of soil fungi, and alert neighbors of developing water stress. Plants release familiar chemicals—scents—that attract pollinators and seed dispersers. In many cases, the particular blend of compounds released can be complex and specific, a way for the plant to communicate its identity to a particular pollination partner. In other cases, generalized scents advertise sweet fruit to a variety of dispersers.

LIVING TOGETHER IN COMMUNITIES

In the mid-twentieth century, the new field of ecology debated whether plants lived together in highly interdependent communities or plant species co-occurred as neighbors of convenience. In 1916, Frederic Clements observed that regions were clearly defined by particular collections of species. Patterns emerged regionally and locally, with specific plant associations seen in spots with specific soil and climate conditions. In northeastern North America where he worked, Clements noted that when plants recolonized abandoned farm fields, they did so with a predictable pattern of successive species. First came grasses and herbs like asters (*Symphyotrichum*), ragweed (*Ambrosia artemisiifolia*), and goldenrod (*Solidago*). These were followed by sun-loving shrubs like blueberry and huckleberry (*Vaccinium*). Easily dispersed, hardy, and short-lived trees like pin cherry (*Prunus pensylvanica*) and aspen (*Populus*) germinated in the sunny meadows and quickly developed a canopy, which shaded the soil, reduced its temperature, and stimulated germination of longer-lived trees like sugar maple (*Acer saccharum*) and beech (*Fagus*). Deeper shade beneath this canopy fostered oaks (*Quercus*) and a variety of ferns. Clements argued that these species depended on each other to provide the conditions needed for germination and growth, so they must live as an interconnected community, vulnerable to the loss of any of its members. Henry Gleason offered a counterargument in 1917. He proposed that individual plant species simply followed their preferred environmental conditions. If particular species tended to be found together, it was because they all sought similar climates, soil types, moisture, and light environments.

Over the next decades, several important observations shifted thinking toward Gleason's view of plant communities. The species that recolonized an area after wildfire or clear-cutting differed depending on the type of disturbance. While grasses and herbs dominated abandoned farm fields, bracken fern greened areas cleared by wildfire. Moreover, paper birch (*Betula papyrifera*), not yellow birch (*B. alleghaniensis*), was the first tree to invade areas blackened by fire. However, the arrival of forest trees also depended more on the proximity of mature trees that could provide seed than on the particular species of the tree. These observations suggested that plant species responded to cues in their environment and the luck of dispersal, rather than following specific partners.

In Wisconsin, John Curtis described the slow dance between prairie and forest. In 1959, he reported that the boundary was dynamic and varied with climate, fire history, and dispersal as species from the forest ventured into the prairie and prairie species penetrated open spots in the forest. Fossil pollen preserved in the sediment of ponds throughout the eastern United States provided additional evidence. In 1958, Margaret B. Davis published her first paper on pollen deposited in ponds during the retreat of glaciers at the end of the most recent ice age. Davis's work documented 21,000 years of forests greening a landscape that had been scraped bare by ice. Over the next 20 years, Davis showed that tree species followed individual migration routes out of their glacial refuges. She realized that tree species living together today were those that happened to arrive at a spot where climate and soil conditions were right.

The final piece of evidence came from ecologists studying how North American forests responded to the chestnut blight of the early twentieth century. American chestnut (*Castanea dentata*) was one of the most abundant trees throughout the Appalachians until the late nineteenth century thanks to forest husbandry by Indigenous Americans. In Clements's view of interconnected plant associations, chestnut's loss would have ripped the forest community apart. Chestnut was abundant in the pollen record prior to the early twentieth century, then it suddenly disappeared as mature trees succumbed to a fungal disease introduced from Asia. By 1912, almost all mature chestnuts in North America were dead or dying. Instead of collapsing, however, other trees like maples and oaks filled the spaces left by the chestnuts. The forest communities changed but did not collapse. The evidence seemed clear that plants followed preferred soil and climate conditions and established wherever their seeds could find a space. Although canopy trees certainly changed the environment on the forest floor and influenced which species could live in their shade, the identity of canopy trees seemed to matter little.

Applying these ideas to the Mesozoic, paleobotanists recognize that plants have relatively simple needs that have remained the same over most of the history of plants on land: sunlight, water, mineral nutrients in soil, and space. The constancy of these needs means that many of the principles of plant biology derived from the study of modern plants apply easily to those that lived in the Mesozoic. And where details might differ, we have Mesozoic survivors—ferns, horsetails, conifers, *Ginkgo*, and cycads—with which to test the breadth of these biological ideas.

GROWTH

The differences between plants and animals begin with their cells. Both plant and animal cells have nuclei that contain most of the genetic instructions—in the form of DNA—needed to build and operate the organism. Plant and animal cells also share structures called ribosomes that translate genetic information into proteins that do the work of life, and mitochondria that use sugars for energy to power all the organism's activities.

A flexible cell membrane encloses all these cellular structures and selectively allows water and other molecules in and out of the cell. This is where the differences begin. Plants have chloroplasts, the factories where photosynthesis takes place. Chloroplasts contain the pigments plants use to gather light, split water molecules, and package chemical energy. Chloroplasts also house the scaffolding where the photosynthetic enzyme rubisco grabs carbon dioxide and begins its transformation into sugar. The most consequential difference between plant and animal cells is the stiff cell wall that surrounds the cell membrane in plants. Plant cell walls are composed mostly of cellulose, the long molecule that gives cotton its string-making powers. Some cell walls, particularly those in water-conducting cells, are further reinforced with a virtually indestructible compound called lignin, which gives wood its woodiness.

The cell wall gives plants many advantages. First, they are stiff and resilient. Plants do not need bones or shells to hold their bodies together and upright on land. Cell walls also protect plant cells from invasion by ever-present microbes and make them a little less palatable to predators. Of course, both microbes and plant predators have evolved strategies to deal with these defenses. For example, one group of fungi found the biochemical key to digest lignin. The protective cell wall also comes at a cost: after they lay down their walls, plant cells cannot move. In animal growth and development, cell movement is vital. Most animal cells are specialized for a particular function—for example, becoming muscle, blood, bone, or a reproductive cell—soon after they arise and must move to the appropriate part of the animal body to function. This strategy will not work for plants. Instead, plants retain pockets of generalized cells throughout their body, allowing them to spring into developmental action and transform into specialized cell types whenever needed. In practical terms, when you prune your shrubs, the plant can activate generalized cells near the site of the cut and immediately build new water-conducting, sugar-transporting, structural, or photosynthetic parts. Most animals, in contrast, struggle to replace important cells when they are lost or damaged.

MODULAR GROWTH IN PLANTS

Plants build their amazing variety of forms from a simple building block, creatively modified and repeated over and over. To understand this design, consider a familiar houseplant like the jade plant (*Crassula ovata*). Look closely at a straight part of the stem. You will notice that leaves emerge, generally in pairs, at defined points along the stem, with successive pairs oriented at right angles to the pair above and below. Between the leaf pairs, the stem is straight and smooth—or perhaps wrinkly and textured in older stems. The point where the leaves emerge is called a node. The straight stem between is an internode. In an unbranched jade plant, the node-internode building blocks stack one on top of the other like children's snap-together bricks to form a straight stem with pairs of leaves all along. Growth begins with a cluster of generalized cells tucked between the topmost pair of leaves. These cells retain the ability to transform into any cell type that the plant needs. The generalized cells at the growing tip produce a plant hormone called auxin, which directs the proliferation of cells and helps the new cells become organs, such as leaves at the edges of the stem or water-conducting cells toward the center. When all the needed module cells are ready, a second family of hormones, gibberellins, signal cells to elongate by filling with water and stretching, just before the stiff cellulose cell wall forms. Once the final shape of the cell is cast by the cell wall, the new cell is ready to begin its work. In a plant instant, a new shoot extends from the growing tip, the straight internode elongates, and leaves unfurl. The

Crassula ovata illustrates the node–internode design found in most land plants. Leaves and branches emerge from nodes. Between nodes—the internode—the stem elongates.

plant has added a new node–internode module, and the leaves at the growing tip have moved just a bit closer to the light. During the phase when gibberellins direct cell elongation, the plant's light-sensing system samples the environment to determine whether the growing tip is in sun or shade. If phytochrome detects shade, gibberellins pour into the cells, and the internode stretches as far as it can in search of better light conditions. If phytochrome detects the red wavelengths of full sunshine, gibberellins fade and the plant produces a short internode.

Take a pair of pruning scissors and cut through the jade plant's internode just above any pair of leaves and you will see another key part of the plant growth strategy. Nestled in the node, just above the place where the leaf attaches, is another cluster of generalized cells, a bud. When growth is focused at the tip of the stem, auxin produced by cells at the top of the plant keeps the buds below in a state of suspended animation. They remain able to develop, but they do not. However, remove the tip of the stem, and the auxin supply disappears. Now, the generalized cells closest to the topmost remaining pair of leaves spring into action and begin to pump out their own auxin. Auxin stimulates development of a new branch—long, straight internode and leafy node—that emerges from the spot just above the topmost remaining leaf. In addition to becoming branches, the generalized cells at each node can become reproductive structures like flowers or cones when the plant senses that the time is right.

Now return to the cutting you removed from the jade plant. Place it in water, and the generalized cells at the node just above the cut will recognize that something dramatic has happened and respond by transforming into roots. Once roots emerge, the new plant can be transferred to soil, where it will continue to grow upward from the shoot tip and downward as roots from the bottommost node. In nature, wind or a passing animal might knock branches off the wild jade plant. Once on the ground, the node closest to the break will send out roots that sense gravity and dive downward in the hope of reestablishing a connection to the lifegiving soil. The shoot tip also senses gravity and turns to grow upward. This ability to respond to damage allows the plant to proliferate in spots where the parent thrives.

BUILDING THE VARIETY OF PLANT FORM WITH NODES AND INTERNODES

Although a jade plant provides a useful example of basic plant growth, many plants look quite different, with less obvious nodes and internodes. Even so, they use the same succession of modules, with a few modifications, to produce their varied architecture. Consider a conifer like eastern white pine (*Pinus strobus*). When a seed germinates, the embryo's root tip emerges first. Starch granules in the root cells sense gravity, and the tip dives for the cover of soil. Meanwhile, the shoot wriggles free from the seed coat and displays a tuft of needles. Phototropins immediately look for the blue light that will point the shoot skyward. Once upright, the cluster of generalized cells nestled within the first lock of needles produces growth-stimulating auxin, and new modules develop. If phytochrome detects shade, the seedling's best chance of survival will be to devote energy to producing as many new leaves as possible. Therefore, successive internodes will be

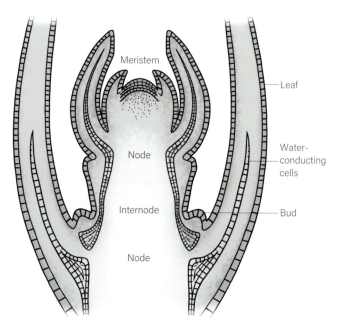

The growing tip or meristem of the plant contains cells that can develop any plant tissue, giving the plant flexibility to replace lost or damaged parts.

extremely short, with needles produced at each node. So that needles do not shade each other, they emerge in a spiral around the cylindrical shoot, creating a bottle-brush array. Because the internodes are short, the seedling does not grow taller during this early phase. Over time, the forest canopy might open, allowing red-enriched sunlight to strike the pine seedling. Phytochrome senses the change and signals gibberellins to stretch the internodes so that the seedling can shoot up quickly. Pines, like most conifers, depend heavily on their topmost growing tip to regulate their shape, and as that growing tip moves farther away from the nodes below, the buds at these lower nodes feel the apical tip's auxin less and begin to stir. When the growing tip is just the right distance from the buds below, they produce new branches. For many conifers, branches arise at regular intervals along the straight, central stem as the tree grows taller. For pines growing in a forest where the only red light is from above, lateral branches may see shade, which triggers growth to stop and the branches to die and fall off. This creates the tall, straight, branchless trunks that characterize conifers in dense forests.

This cross section shows how cycads build sturdy trunks. Their strength comes from a thick outer bark and woody leaf bases. Cycads produce little or no wood.

Some ferns and grasses take a different approach to nodes and internodes. The green clumps of ferns or the grass blades between our toes are only the leaves of the plant. Dig down a bit and you will find the leaf clusters, produced on very short internodes, attached to a specialized stem that grows horizontally just below the soil surface. For the underground stem, the dark of the soil environment stimulates the internodes to elongate as the growing tip explores for soil nutrients or a patch of light. Once the growing tip detects a suitable spot, gibberellin production slows, internodes shorten, buds spring into action, and a new tuft of leaves grows skyward. For these plants, their growing tips and the generalized cells that control plant growth are hidden safely below the surface of the soil, protected from animals who gobble up the leaves.

To build even more exotic plant bodies, evolution has modified the basic module even further. In the squat trunks of many modern cycads, internodes fail to elongate, and whorls of leaves and the woody scales that support them pile on top of one another to form a spiky mound. Precious generalized cells are protected inside the armored trunk while only leaves face a world full of herbivores. Vines like kudzu (*Pueraria montana*) take the opposite approach. The fast-growing invasive from subtropical regions of South and East Asia can grow up to 30 cm (12 in) a day by maximizing cell elongation in its internodes. Vines combine this extraordinary elongation ability with their light-sensing and support-grabbing powers to locate sunlight and climb toward it.

MANY WAYS TO BUILD A TREE

Imagine taking your childhood sweetheart to a favorite tree to enjoy a first kiss. You might—but really shouldn't—commemorate that first love by carving your initials into the tree's bark. Decades later when you return to reminisce, what will you see? The letters, now broad and distorted, remain at child's height, even though the tree is now much taller than it was when you visited long ago. The increase in trunk diameter is only one of the ways plants become trees sturdy enough to withstand storm winds and hold up a canopy full of leaves.

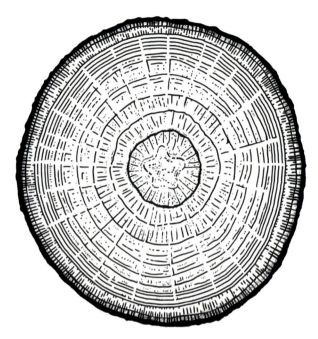

Many seed plants produce woody trunks with a thin layer of growing cells that proliferates just under the bark in this cross section. In seasonal climates, alternating periods of growth and dormancy produce rings.

Tree size has evolved in all groups of land plants that have tracheids to conduct water. When similar traits are seen evolving repeatedly in different groups, it is a good bet that the trait offers a significant benefit for the lineage that possesses it. For plants, light is a vital resource, and it only comes from above. Mesozoic plants found several solutions to the challenge of becoming tall.

The most common way to become a tree is to make wood. Wood is simply abandoned water-conducting cells. In conifers, ginkgos, some cycads, and many closely related extinct groups from the Mesozoic, the tree maintains a layer of generalized cells just under their bark. These cells, called the cambium, divide, with the cells toward the outside of the layer becoming the sugar-transporting phloem cells, and the cells on the inside becoming the water-conducting cells, or xylem. In seasonal climates, cambium cells begin dividing at the beginning of the growing season, creating a new layer of large, water-conducting cells that jump-start leaf expansion in the canopy high above. Cambium cells continue to divide throughout the growing season until they quiet as the tree becomes dormant. This pattern of activity and dormancy creates the familiar tree rings

Palms, a group of flowering plants, build sturdy trunks with a combination of thick woody bark and sturdy fibers that run up and down the trunk providing strength and flexibility.

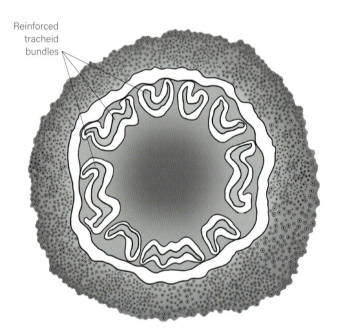

Tree ferns mantle their stems in fibrous roots that give the trunk strength and flexibility. Within the stem itself, bundles of water-conducting cells are reinforced to make the trunk stiff.

that characterize wood formed in seasonal climates—in both warm–cold and wet–dry seasons. Trees that experience a year-round growing season continually make new conducting cells and do not produce rings. Making new wood and the resulting increase in trunk diameter helps the tree in two ways. First, growth replaces water-conducting cells that no longer function because their water chain has been broken. Second, continually adding wood means that the tree's trunk grows stronger as its canopy enlarges and becomes heavier.

Most angiosperms build trees with a cambium, a skill they inherited from a common ancestor with the other seed plants. However, palms evolved a different approach. Instead of a cambium layer, palms cluster all their growing cells at the top of the plant where the leaves emerge. The palm stem contains bundles of water-conducting cells scattered throughout the trunk rather than ringed wood. The multitude of conducting bundles, combined with lignin-reinforced fibers, give the palm trunk the strength and flexibility to withstand tropical cyclones. Palm seedlings grow by first expanding laterally to achieve their adult diameter. This produces a low, squat trunk topped with a shock of leaves. This early phase requires a lot of energy and explains why tree-sized palms prefer sunny spots. In contrast, shade-loving palms generally remain small

and slender. After widening, the generalized cells of the palm heart continue to divide and elongate but shift toward upward growth. Also, because palms cannot replenish their water-conducting cells with new growth from a cambium, they cannot recover when the water in their conducting cells freezes. Consequently, palms have reliably indicated warm climate for their 85-million-year history.

Ferns also lack a cambium and those that become trees use a strategy like that of palms. Tree fern trunks have complex, netlike webs of thick-walled conducting cells, lignin-reinforced, rodlike fibers, and sturdy leaf bases that remain on the trunk to provide extra support. Many tree ferns also swaddle their stems in a mantle of special roots that add more support. Ferns have another trick to help keep their leaves in the canopy as their trunk grows taller: they extend their leaves on the plant equivalent of a selfie stick. As a tree fern's growing cells begin to sense shade, phytochrome turns on the gibberellins in the leaf's petiole to extend it from a modest 50 cm (about 1.5 ft) to more than 300 cm (over 9 ft), simply by making individual cells longer. This allows the leaves to reach the encroaching canopy even if the trunk cannot grow fast enough. Eventually, however, the tree fern may be overtopped by faster-growing neighbors and adjust to life in the shade.

EVOLVING IN RESPONSE TO THE ENVIRONMENT

HOW EVOLUTION HAPPENS

Genetic Variation and Natural Selection

Evolution means changes in a lineage that happen over many generations. In addition to time, evolution requires individuals to differ from one another genetically. Like animals, plants generate genetic variation through mutations—changes in the DNA sequence—that happen when cells copy their genes as they divide.

Because animal cells specialize very early in development, the only mutations that travel into the next generation are those that happen in egg cells and sperm. Mutations in any other part of the body perish with the animal. Plants have more possibilities. Because each of a plant's growing tips retains generalized cells that can become eggs and sperm, mutations that happen in any of a plant's many growing tips can be passed down to the next generation. For large, long-lived trees, the thousands of growing tips in a single canopy may each contain a different suite of mutations that can become

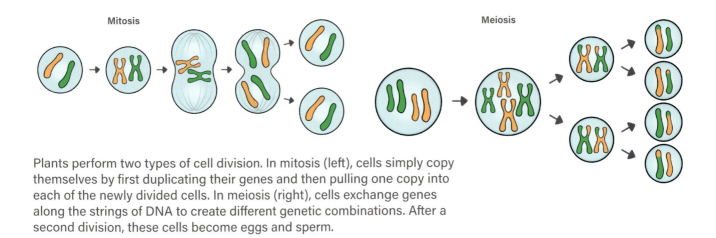

Mitosis

Meiosis

Plants perform two types of cell division. In mitosis (left), cells simply copy themselves by first duplicating their genes and then pulling one copy into each of the newly divided cells. In meiosis (right), cells exchange genes along the strings of DNA to create different genetic combinations. After a second division, these cells become eggs and sperm.

part of the tree's evolutionary heritage. Next, sexual reproduction adds to genetic variability by shuffling the genes of the parents to produce new combinations in offspring. Plants and animals both do this, but plants have a few additional tricks for amplifying the variation generated by sex. Plants can combine and multiply all of their genes at once—a catastrophic genetic "error" that would be fatal to most animals. This allows plants to combine more traits and generate new species with novel features. The duplication of the plant's

genes that happens during these events can also boost useful traits. For example, a number of weeds have evolved resistance to common commercial herbicides by simply adding more copies of the gene targeted by

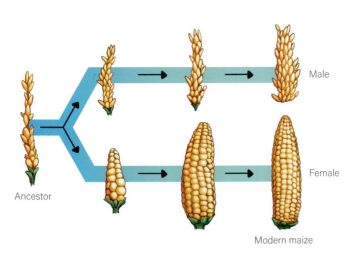

Male

Female

Ancestor

Modern maize

An ancestor of modern maize (above left) produced separate pollen- and seed-producing flowers that looked similar. Over time, farmers chose those individuals that made larger ears with more grain.

RIGHT: Wild teosinte grows in branching clumps. This species contributed genes to domestic maize.

the chemicals. Generating tremendous genetic variety quickly allows plants to evolve rapidly.

The environment tests the traits produced by genetic variability through a process that Charles Darwin called "natural selection." In developing this idea, Darwin riffed on a practice that farmers have used for millennia: they chose traits that they liked and encouraged individuals possessing those traits to reproduce. Over generations, breeders could sculpt plants and animals with the traits they preferred. The evolution of domestic maize (*Zea mays*), also known as corn, illustrates the process. A wild grass that is likely extinct today produced a few popcorn-like seeds that caught the attention of Indigenous Americans in what is today central Mexico. They began to save seeds and plant them in their gardens. However, the ancestral plant was not particularly productive. In a chance event about 5,000 years ago, pollen from a related wild species accidentally combined with early maize and brought with it new genes that produced a compact cob with many more kernels. An observant Indigenous farmer recognized the potential in the new combination and began to grow it, choosing the largest ears in successive generations. Over time, selective breeding of individuals with large and nutritious seeds produced the crop that became a staple throughout the Americas. Darwin proposed that the environment governed which organisms produced offspring, much like a farmer selected traits. Over long stretches of time, this process could produce all the variety observed in the biological world.

Evolution for Flexibility

In animals, new species commonly evolve as organisms specialize to share key resources. Plants have simple needs—light, water, a handful of common minerals, carbon dioxide, and space to grow—that cannot be easily divided to generate new species. Instead, new plant species explore different strategies for acquiring a limited resource. Light, for example, generally comes from above, so any trait that allows plants to move skyward, whether through climbing or growth upward on sturdy branching stems, will be favored by natural selection. Similarly, plants are rooted in place throughout their lifetimes. Therefore, selection favors strategies for flexibility in changing conditions, rather than increasing specialization. For example, most plants can adjust the configuration of their leaves to accommodate differing light conditions throughout their lifetime. For instance, plants that live on the forest floor experience shade until a gap in the canopy opens above them.

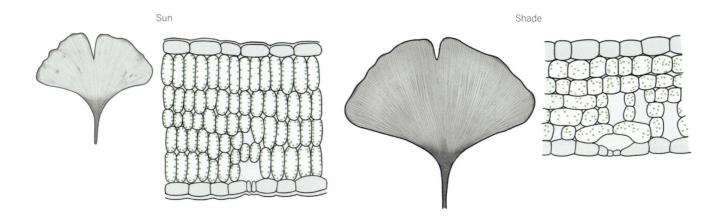

Sun   Shade

Leaves of modern *Ginkgo biloba* demonstrate that leaves developing in full sun (left) are smaller and, in microscopic section, thicker with elongated photosynthetic cells. Leaves grown in the shade, even on the same tree (right), are larger and thinner.

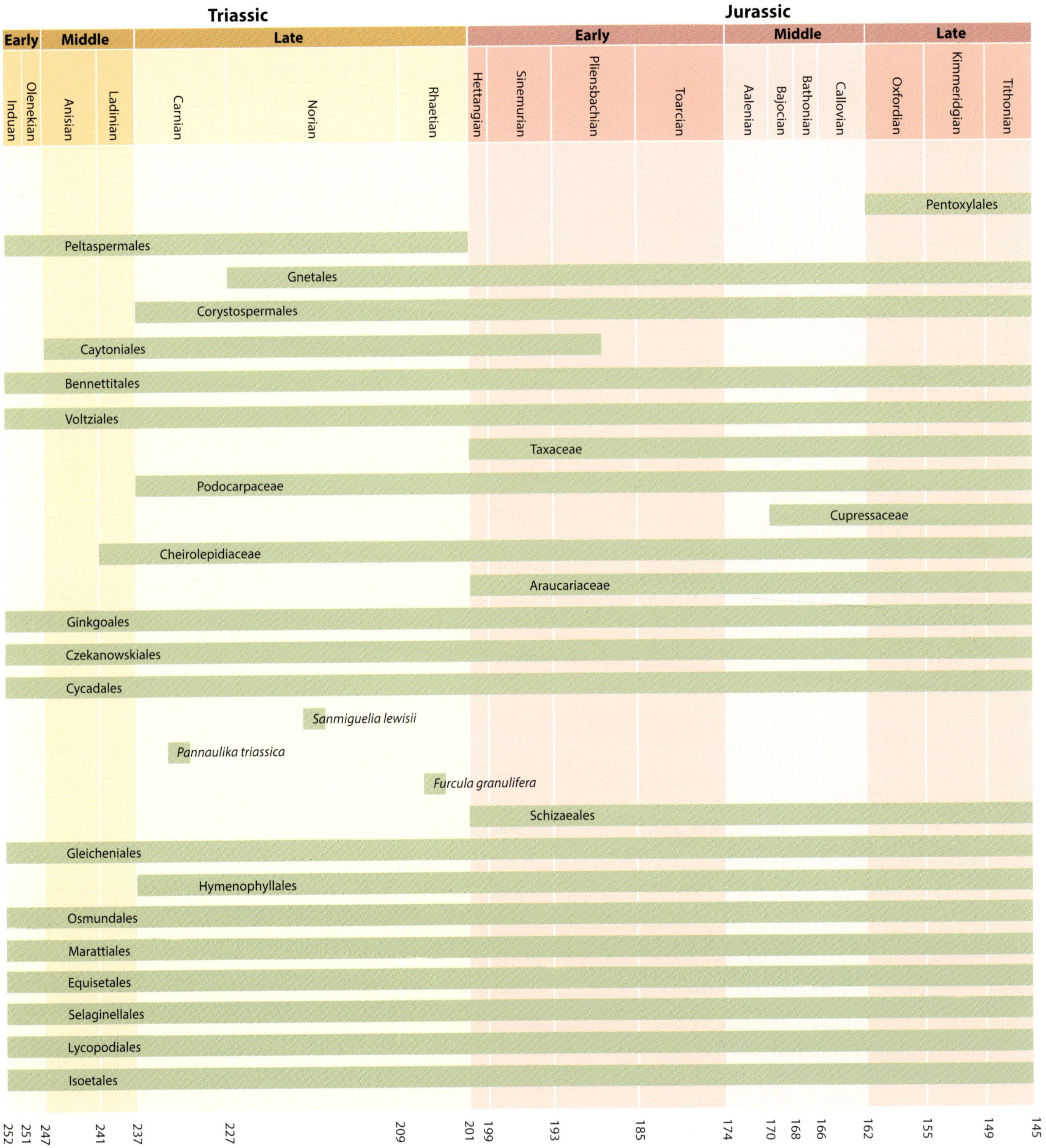

Triassic

Jurassic

| Early | Middle | Late | Early | Middle | Late |

Induan · Olenekian · Anisian · Ladinian · Carnian · Norian · Rhaetian · Hettangian · Sinemurian · Pliensbachian · Toarcian · Aalenian · Bajocian · Bathonian · Callovian · Oxfordian · Kimmeridgian · Tithonian

Pentoxylales

Peltaspermales

Gnetales

Corystospermales

Caytoniales

Bennettitales

Voltziales

Taxaceae

Podocarpaceae

Cupressaceae

Cheirolepidiaceae

Araucariaceae

Ginkgoales

Czekanowskiales

Cycadales

Sanmiguelia lewisii

Pannaulika triassica

Furcula granulifera

Schizaeales

Gleicheniales

Hymenophyllales

Osmundales

Marattiales

Equisetales

Selaginellales

Lycopodiales

Isoetales

252 · 251 · 247 · 241 · 237 · 227 · 209 · 201 · 199 · 193 · 185 · 174 · 170 · 168 · 166 · 162 · 155 · 149 · 145

Time ranges for major groups discussed in the text. The dotted line indicates that the group was absent from the fossil record for a time but not yet extinct.

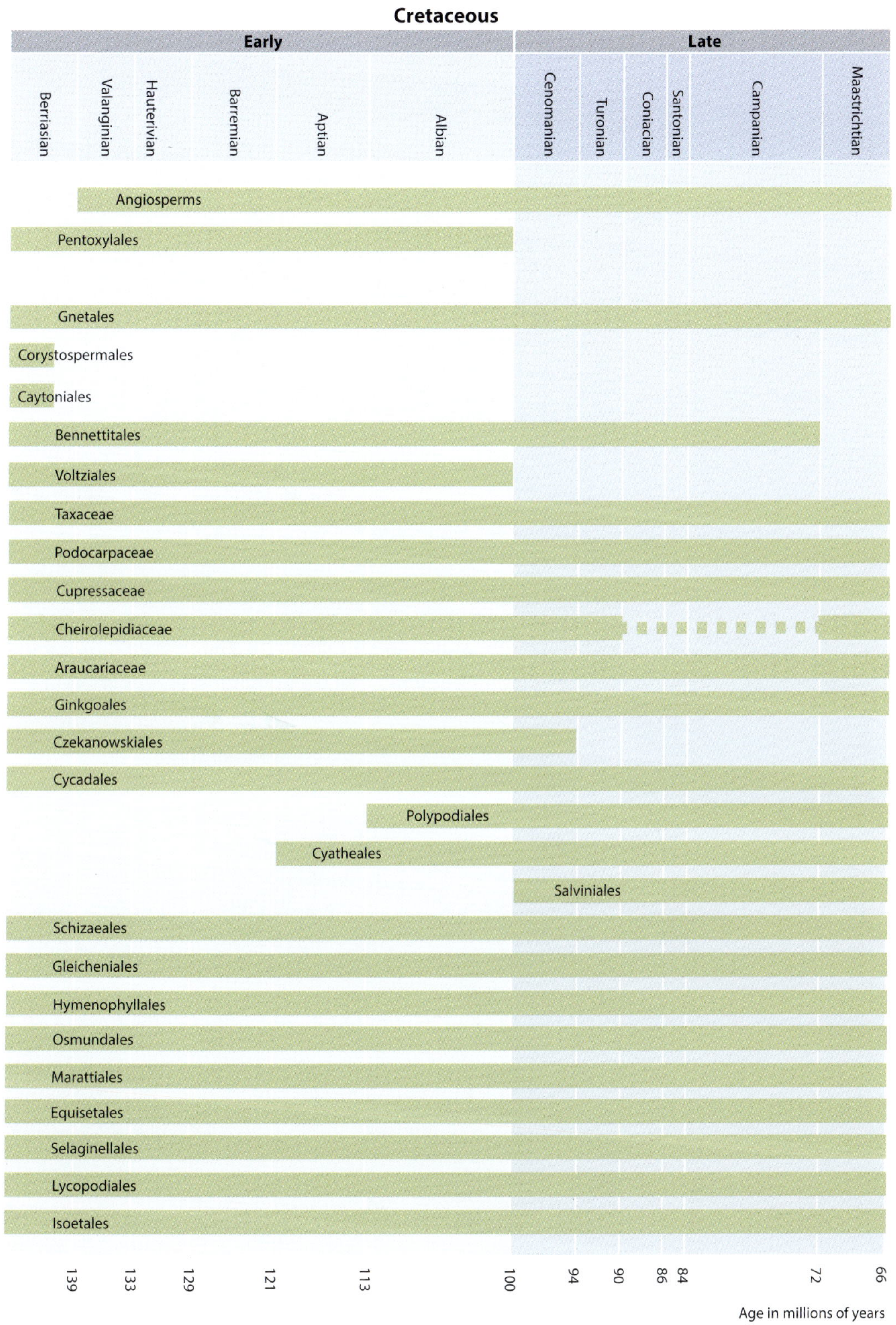

Cretaceous

Early						Late					
Berriasian	Valanginian	Hauterivian	Barremian	Aptian	Albian	Cenomanian	Turonian	Coniacian	Santonian	Campanian	Maastrichtian

Angiosperms

Pentoxylales

Gnetales

Corystospermales

Caytoniales

Bennettitales

Voltziales

Taxaceae

Podocarpaceae

Cupressaceae

Cheirolepidiaceae

Araucariaceae

Ginkgoales

Czekanowskiales

Cycadales

Polypodiales

Cyatheales

Salviniales

Schizaeales

Gleicheniales

Hymenophyllales

Osmundales

Marattiales

Equisetales

Selaginellales

Lycopodiales

Isoetales

139 133 129 121 113 100 94 90 86 84 72 66

Age in millions of years

Now intense sunlight floods the forest floor, damaging delicate chloroplasts in leaves tuned for shade. Most plants have the ability to shed these damaged leaves and grow new ones calibrated for their new light environment. When the canopy closes again and the plant finds itself back in shade, it will shed its sun leaves and make a new set of shade leaves. Neither the sun- nor shade-type leaves have been specifically selected by this patchy environment. Instead, plants that had flexibility to change with the environment were more likely to survive longer and reproduce more. Evolution favored resilience in a constantly changing environment. The simplicity of a plant's environmental needs also means that many pathways lead to survival. For example, if rain is rare, some plants evolve strategies for storing water, others conserve it, and others develop seed dormancy in which they germinate, grow, and reproduce quickly in the few weeks following a rainstorm.

All these strategies will allow their bearers to successfully weather drought and send offspring into the next generation—the definition of evolutionary success.

Although the rule in plant evolution is flexibility, angiosperms provide an exception. One key to the tremendous diversity of flowering plants is the complex relationships with pollinators and seed dispersers. Flowers evolved colors, shapes, and scents to attract specific pollinators because a plant produces the most seeds when pollinators only visit other members of its species. In this case, the advantages of specificity are worth extra energy to the plant. A yucca (*Yucca filamentosa*) produces a scent attractive only to a single species of yucca moth (*Tegeticula yuccasella*) and times its flowering to the season when adult moths emerge and begin to look for mates. Yucca moths seek out the flower and deposit a few eggs within, picking up pollen as they do. They then seek out another flower of the same species to lay more eggs, spreading pollen as they go. When the moth larvae hatch, they eat a few of the developing seeds, a small price for the plant in exchange for having pollen faithfully delivered. Flowering plants have also evolved to cater to the tastes of specific dispersers. Apples, for example, evolved first in a dispersal partnership with birds in what is today China. These early apples were bright red, a color birds easily recognized amid a sea of green foliage, with small, hard, bitter fruit that suited birds that swallowed the fruit whole. As the lineage migrated into central Asia, it left its bird partners behind and developed dispersal relationships with mammals. Mammals have a notorious sweet tooth and were looking for more substantial meals. The new relationship favored trees that produced the large, sweet fruit that we enjoy today. In these cases, angiosperms adopted the animal pattern of dividing recourses—pollinators and dispersers—among many different species to multiply their own diversity and become the most varied lineage of plants in Earth's history.

Yucca filamentosa has evolved an exclusive pollination relationship with a single species of tiny moth, *Tegeticula yuccasella*.

ENVIRONMENTAL FORCES SHAPING PLANTS

Water and the Climate that Delivers It

To plants, temperature is simply an index of water availability. In a cold season, for example, ice crystals form in conducting cells, disrupting the chain of water molecules on the way to the leaves. The leaves experience the loss of water, not the cold directly. Similarly, in a hot, dry season, leaves open their pores to allow water to escape, taking excess heat with it. This pulls hard on the column of water molecules along the stem and may break the chain. Once again, the leaves experience a water shortage. In both hot and cold climates, when the change is seasonal, long-lived plants evolved the ability to sense the oncoming change and drop their leaves before water runs short. By acting preemptively, the tree can harvest valuable nutrients like nitrogen, potassium, and magnesium from leaves before they are shed and store them in the roots until favorable growing conditions return.

Plants have evolved other features to thrive in a range of climates. In rainy regions, the leaves of flowering plants have pointed tips that help shed accumulated water so that it does not clog pores and restrict airflow into and out of the leaf. In dry climates, small, chunky leaves with a thick, waterproof covering retain limited moisture. In the early 1900s, Irving Bailey and Edmund Sinnott observed that broadleaf trees in cool climates tended to produce leaves with serrated edges. Researchers do not understand how this feature aids the tree's survival, but even if the function of all the plant's features cannot yet be explained, there is no question that plants are exquisitely adapted to the climate where they grow. Wladimir Köppen's global climate classification scheme published in 1884 was fundamentally a mapping of vegetation types. Later, when meteorological measurements became available worldwide, the vegetation zones correlated almost perfectly with temperature, precipitation, and seasonality patterns.

Disturbance

Environmental disturbance was ubiquitous in the Mesozoic. Warmer climates contributed to frequent wildfire, even in a time when atmospheric oxygen levels were somewhat lower than those of today. Warm climates also intensified storms that shattered limbs and downed trees. Triassic monsoons in what is today the western United States generated floods that swept away hundreds of trees at a time. And during the Jurassic and Cretaceous, large dinosaurs crashed through the vegetation. Unable to run or hide, plants were burnt, uprooted, swept away, or crushed by a myriad of Mesozoic disasters. And in addition to the everyday disturbances of Mesozoic life, plants passed through three animal mass-extinction events that were each characterized by catastrophic disruption of the physical environment. Even if extinction events did not wipe out as many groups of plants as animals, the conditions responsible for widespread extinction certainly killed uncountable individual plants and disrupted their communities.

Disturbance acted as a powerful selective agent, killing many individual plants. Those that survived, and thus passed on their traits to subsequent generations, possessed a variety of features that helped them cope with disturbance. Plants adapted to fire shelter pockets of generalized cells safely below the soil surface. When aboveground stems burn, these buds activate and grow new shoots. Other plants protect seeds in sturdy cones that remain clenched until the wildfire's heat melts resins on the cones' surface and allows them to open. For these plants, the fire also removes organic matter and competitors from the soil surface, making a perfect bed for the seeds that drop from the newly opened cones. Other species make seeds that can persist in the soil for long periods and germinate when their phytochrome senses detect that the canopy has been removed. Other trees like the coast redwood (*Sequoia sempervirens*) evolved thick, fire-resistant bark to insulate their delicate living cells from the heat of a blaze. *Sequoia* also sheds most of its low branches as it reaches its full height so that wildfire cannot climb into the canopy and destroy vulnerable foliage.

Many of the innovations that make flowering plants so successful may have begun as adaptations to disturbance. The high angiosperm photosynthetic rate may have its roots in quick responses to a changing light environment in their ancestral forest floor habitat. Likewise, fast growth and reproduction may have allowed flowering plants, germinating on the banks of frequently flooded rivers, to complete their life cycles before they were swept away. This ability allowed angiosperms to occupy a piece of the landscape unavailable to slower-growing species. In a world with limited space to grow, this strategy helped angiosperms leave more offspring in the next generation. Fast growth also helped flowering plants recover from trampling and colonize ground churned by passing dinosaurs, conditions that slower-growing seed plants could not exploit.

Predators, Pathogens, Partners, and Dispersers

For animals like us, plants are resources—food, shelter, and, perhaps, beauty. Plants, on the other hand, have evolved a myriad of ways to coax animals to work for them. Some plants surround their seeds in sweet and nutritious fruit to encourage animals to consume and transport their seeds, depositing them in a pile of ready-made fertilizer. Plants advertise such seeds and fruits with colors and scents tuned to the animal partner's senses. Other plants evolved seeds with hooks to catch fur, skin, and feathers for a free ride to a new spot. Plants also use some of their surplus sugar to offer sweet rewards for animals who visit their reproductive structures and carry pollen from plant to plant. And to encourage fidelity, natural selection tuned the colors and scents to the preferences of a specific set of animal partners. For example, the flower of the hammer orchid (*Drakaea*) smells exactly like a female thynnine wasp. The male lands on the flower and becomes covered with pollen as he attempts to mate. He then flies off in search of a new partner, carrying pollen to the next orchid as he goes.

Plants cannot evade predators by running or hiding. Instead, they protect themselves with physical and chemical defenses. The Mesozoic introduced many new plant predators to the evolutionary stage: large (herbivorous dinosaurs), small (voracious mammals), and minuscule (several new groups of insects). Some lineages like conifers and cycads evolved armor—investing their

long-lived leaves with the sturdy molecule lignin and imbuing them with sticky and bitter resins. Although this challenged delicate mouths and sensitive tummies, many large dinosaurs, particularly the long-necked sauropods, seem to have specialized on tough foliage. They may have evolved a style of digestion similar to that of modern cows, sheep, goats, deer, and camels, which consume indigestible plant material and let the microbes in their gut do the biochemical work. Some paleontologists speculate that the largest herbivorous dinosaurs may have done the same.

Plants have also evolved a wide variety of biologically active molecules that they use to defend against predators. Most of these chemicals began in biochemical pathways evolved to deal with plant waste products. Since plants lack a liver, kidneys, and lymphatic system to remove waste products from their cells, they eliminate wastes at the cellular level by making unwanted or toxic molecules into something harmless to the plant. When left in leaves, some of these molecules are toxic (or just bad-tasting) to the animals that sample them. Plants with the widest range of such chemicals persisted in the predator-filled Mesozoic forests. Over time, evolution sculpted these defensive chemicals to mimic biochemical pathways in the animals that consumed them. For example, caffeine, a compound made by a number of flowering plants, stunts the growth of insects that consume the leaves or fruits that contain it. Another molecule stashed under the bark of willows (*Salix*) offers antimicrobial protection for the tree, and for humans, who use it in traditional medicine and in Western medicine as a pain reliever and in over-the-counter acne treatments. Plants produce chemicals that, when consumed by animals, stimulate appetite, control blood sugar, stop or start the heart, kill cancer cells, alter brain function, and much, much more.

Plants evolved molecules to signal neighboring plants and other partners. At the most basic level, plants produce chemicals to entice pollinators and seed dispersers. Other sophisticated chemical signals attract soil fungi. Fungus and root each release specialized chemicals that stimulate mutual growth and interconnection. The fungus releases defensive chemicals that ward off parasites and helps the plant gather water and minerals from the soil. The plant reciprocates with sugary food for the fungus. Some call these interconnections the "wood wide web" because the underground network of roots and fungal cells seem to transmit information through the soil. Plants starved for light send chemical signals through the fungal network to larger neighbors that then coax fungi to share sugar, with the fungus taking a bit for its trouble. Although science is just beginning to explore these subsurface connections, early research leaves little doubt that in healthy soil, plants, fungi, and a variety of soil microbes form complex, dynamic partnerships in which all parties benefit. Plants also use their communication chemicals to signal one another. When plants are attacked by a herbivore, they emit chemical signals that waft through the air and trigger neighbors to produce defensive chemicals. Although plants do not act with altruistic intent, sensing and responding to an antipredator signal would enhance reproductive success and therefore persist in a predator-filled world.

MESOZOIC CLIMATE SHAPES PLANTS

EARTH'S ATMOSPHERE AND CLIMATE EQUILIBRIA

The Mesozoic climate fluctuated with the quantity of so-called greenhouse gases in the atmosphere. Greenhouse gases take their name from their ability to act like the glass in the windows of a greenhouse. On a summer's day, the sun's visible rays shine through transparent windows and heat up surfaces within. Warm surfaces radiate energy in infrared wavelengths—heat—that raise the temperature of air inside the greenhouse. The same happens with Earth's atmosphere. Visible light speeds through the atmosphere to warm Earth's surface. Warm surfaces radiate heat that raises the temperature of the air around them. Some heat escapes back to space, but some gases in the atmosphere—mainly

water vapor, carbon dioxide, and methane—act like closed windows, preventing heat from escaping. This property of our atmosphere makes Earth habitable. Without the atmospheric greenhouse, Earth's average surface temperature would be about –18°C (0°F), too cold for liquid water and most life. Earth's vast oceans mean that water vapor will always be abundant in the

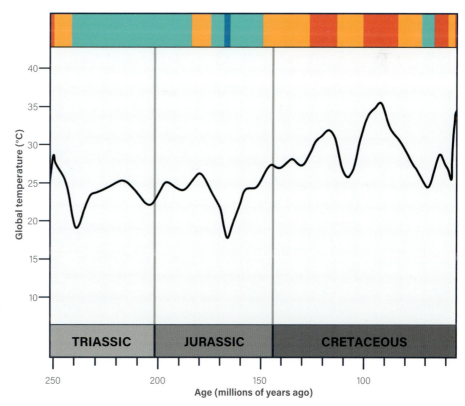

Mesozoic temperature history with icehouse (dark blue), coolhouse (light blue), warmhouse (orange), and hothouse (red) equilibrium states marked. In 2025, the global average temperature was 15°C.

atmosphere. Other greenhouse gases, like carbon dioxide and methane, have varied considerably, and these changes drive Earth between its four stable climate equilibria: hothouse, warmhouse, coolhouse, and icehouse.

Hothouse climates are the warmest and least stable of the climate equilibria. In the hothouse climate, Earth lacked polar ice, and global average temperature may have been more than 10°C (18°F) warmer than today. So-called hyperthermals also characterized the hothouse. During hyperthermals temperature rose rapidly—geologically speaking—to inhospitable levels in response to small changes in greenhouse gases. These geologically sudden heat waves may have lasted tens of thousands of years and devastated some biological communities. Organisms were pushed past their heat tolerances, resulting in mass migration, pest and pathogen plagues, local die-off, or, in extreme cases, extinction.

Warmhouse climate is the most stable climate equilibrium and prevailed for most of the Mesozoic. The warmhouse world also lacked persistent polar ice as global average temperatures hovered 5°C (9°F) or more above today's values. During the warmhouse, forests grew at polar latitudes, and although the poles experienced a freezing season, snow melted with spring and climate remained temperate. Warmhouse climates were generally humid, with ample rainfall to support lush forests.

Coolhouse climate featured the development of polar ice that persisted all year. Snow-white surfaces reflected the sun's energy back to space and further cooled the poles. While tropical temperatures stayed steady, colder polar conditions created a large temperature difference between equator and poles that dried climate in the middle and high latitudes.

The icehouse climate equilibrium was marked by thick, persistent ice sheets at the poles that periodically advanced into the middle latitudes. These so-called "ice ages" characterized the last two million years or so of Earth's history, with the advance and retreat of continental glaciers driven by minute variations in Earth's orbit and amplified by changes in snow cover and greenhouse gases. Although the waxing and waning of continental ice created wild swings in climate that triggered widespread migration of plants and animals on land, these fluctuations happened at regular intervals and over timescales that allowed ecosystems to respond. So, unlike climate instability in hothouse hyperthermals, icehouses produced relatively few extinctions.

ESTIMATING CARBON DIOXIDE LEVELS IN THE MESOZOIC ATMOSPHERE

Over its long history, Earth has moved among these climate states in response to changes in greenhouse gases, primarily carbon dioxide. Despite the importance of carbon dioxide in the climate equation, geologists struggle to estimate how much of the gas was present in the atmosphere of the past. Scientists can directly sample the atmosphere of the last 800,000 years or so by measuring gases trapped in air bubbles in glacial ice. Beyond this, geologists rely on a variety of indirect methods that measure something correlated with atmospheric carbon dioxide. For example, the chemistry of some minerals in soil

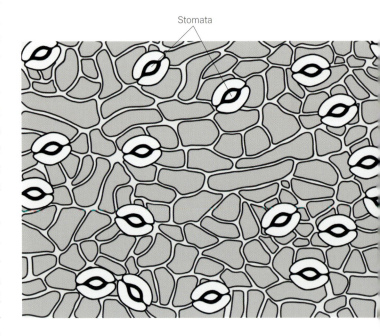

Stomata

Gas exchange pores, stomata, on the underside of a Mesozoic *Ginkgo* leaf reveal the history of carbon dioxide in Earth's atmosphere. When carbon dioxide levels are high, plants produce fewer stomata.

tracks atmospheric carbon dioxide. Geologists also estimate the amount of carbon dioxide emitted from volcanoes and locked up in coal and other sedimentary rocks to model changes in carbon dioxide through time. Paleobotanists recognized another correlation between the number of stomata produced by some plants and atmospheric carbon dioxide. Stomata, specialized leaf pores that control gases moving into and out of the leaf, allow carbon dioxide into the leaf for photosynthesis, so plants have evolved to adjust their number to prevailing conditions. During times of high carbon dioxide, plants make leaves with fewer stomata to allow sufficient carbon dioxide in while minimizing the amount of water that escapes. When carbon dioxide is low, plants need more stomata to allow sufficient carbon dioxide in. However, none of these estimates are precise, and approximations of Mesozoic carbon dioxide vary widely. Nonetheless, a general picture of Mesozoic climate has emerged.

CARBON DIOXIDE LEVELS TRANSLATE TO CLIMATE

At the beginning of the Triassic, carbon dioxide may have been as low as 420 ppmv, a value on par with early-twenty-first-century levels. Geological indicators suggest that during the Early Triassic, Earth was firmly gripped by a warmhouse climate and there is no evidence for polar ice. Carbon dioxide rose steadily through the Triassic and into the Early Jurassic, when levels likely reached 1,000 ppmv or higher. A drop in deep-ocean oxygen concentrations in the Early Jurassic Toarcian age suggests a geologically brief period, when carbon dioxide may have exceeded 2,000 ppmv.

Carbon dioxide levels fell through the Middle and Late Jurassic. Climate shifted from warmhouse to coolhouse, allowing polar and mountain ice to accumulate before the return to a stable coolhouse. Estimates suggest that carbon dioxide levels remained near 1,000 ppmv through much of the Cretaceous, with geologically brief intervals of much higher temperature about 120 million years ago during the Aptian, 105 million years ago during the Albian, and 100 and 94 million years ago during the Cenomanian. These temperature spikes may have been driven by a rapid release

of methane from seafloor sediment. By the end of the Cretaceous, carbon dioxide levels settled at about 700 ppmv, comfortably in the warmhouse.

Most climate indicators agree that the Mesozoic was generally warmer than today, with much lower temperature differences between equator and poles. Seasonality in both temperature and rainfall was also much less than today, with most of the world, except for the interiors of the great continent of Pangaea and then Gondwana, receiving ample precipitation that supported lush forests that produced extensive mid-latitude coal deposits, which are today used for fuels. Moreover, the moisture that these trees pulled from the soil and released through their stomata helped to maintain humid conditions. As forests do today, Mesozoic trees modified and moderated the global climate.

COMPLEX CONTROLS ON GREENHOUSE GASES

Over long stretches of geological time, three processes control atmospheric carbon dioxide: the amount of the gas added by volcanic activity, the amount removed by rock and soil weathering, and the amount stored in rock as fossil fuel and other forms of geological carbon. Most people think of charismatically explosive mountains like Mount Vesuvius or Mount Saint Helens when they think about volcanoes. While eruptions from volcanoes like these can influence climate by adding cooling particles to the upper atmosphere, they contribute little carbon dioxide. Most volcanic carbon dioxide comes from the mid-ocean ridges that bisect many ocean basins. These are zones where Earth's crust pulls apart and new seafloor forms as lava spills out from great rifts. The activity of these ridges varies through time, and the Mesozoic was a particularly active time. Also, a series of especially large lava outpourings, so-called large igneous provinces, punctuated the Mesozoic. These events erupted millions of cubic kilometers of lava and with it came an array of gases, including carbon dioxide. For example, the Siberian Traps at the dawn of the Mesozoic erupted 1–2 million cubic kilometers (240,000–480,000 cubic miles) of lava. The Central Atlantic event during the Early Jurassic contributed

2–3 million cubic kilometers (480,000–720,000 cubic miles) of lava. The Ontong Java Plateau spewed more than 8 million cubic kilometers (1.9 million cubic miles) of lava during the Aptian age of the Early Cretaceous. And the Deccan Traps closed the Mesozoic with another million cubic kilometers (240,000 cubic miles) of lava. These episodes corresponded roughly with intervals of high carbon dioxide and warming climate.

Weathering is the chemical breakdown of rock and soil on Earth's surface. Most weathering occurs in slightly acidic water produced when carbon dioxide in the atmosphere dissolves in raindrops. When the atmosphere contains more carbon dioxide, drops naturally become a bit more acidic. When they fall on rocks, chemical reactions transform minerals in rock into the mineral nutrients plants need, such as calcium, potassium, and phosphorus. In the process, the carbon dioxide transforms from a gas in the atmosphere to minerals dissolved in water. In this way, weathering removes carbon dioxide gas from the atmosphere. These reactions occur faster when carbon dioxide is abundant, conditions are warm, and mountain-building has exposed fresh rock to the rain.

On shorter timescales, the photosynthetic work of plants also modulates atmospheric carbon dioxide. Photosynthesis takes carbon dioxide from the atmosphere and transforms it into organic molecules that become the plant's body. In the process, plants release oxygen. In the normal cycle of photosynthesis and decomposition, which consumes oxygen and releases carbon dioxide, the system remains balanced. However, if organic matter produced by photosynthesis is buried in sedimentary rocks, the associated oxygen stays in the atmosphere. Likewise, if organic matter is removed from Earth's crust and burned, that carbon dioxide returns to the atmosphere and a corresponding amount of oxygen is removed.

OXYGEN, TEMPERATURE, AND WILDFIRE

Plants require oxygen for respiration just as animals do. However, plants produce their own oxygen as they split water during the light reactions of photosynthesis. Therefore, plants are much less sensitive to the amount of oxygen in the atmosphere than are animals, with two exceptions. The first involves photosynthesis itself. The enzyme responsible for capturing carbon dioxide during photosynthesis can also capture oxygen. When oxygen binds with this enzyme, a junk molecule results. Plants recycle this molecule but that takes energy, making photosynthesis less efficient under high-oxygen conditions. Since most estimates suggest that atmospheric oxygen during the Mesozoic was a bit to substantially lower than today's 21%, plants benefited from a boost in photosynthetic performance.

The amount of oxygen in the atmosphere also influenced the frequency and intensity of wildfire, an important disturbance affecting plant communities. Lightning provided ample ignition throughout all of Earth's history. Relatively low atmospheric oxygen rendered fire infrequent throughout the Triassic and most of the Jurassic. However, as oxygen levels and temperatures crept up in the Late Jurassic and Cretaceous, wildfire became more common, coincidentally at the time flowering plants probably first appeared.

CLIMATE, WILDFIRE, AND THE FLOWERING PLANT REVOLUTION

Earth changed dramatically as flowering plants diversified and began to invade existing communities during the Early Cretaceous. The earliest members of the angiosperm lineage grew in the understory of wet tropical forests. Broad swaths of wet lowland tropical and subtropical forest growing during the Jurassic provided ample ground for these first flowering plants. In addition, brief bursts of hothouse heat in the Early and mid-Cretaceous may have helped angiosperms spread. Rapid warming put significant stress on the conifers that dominated the canopies of mid- and high-latitude forests during the Cretaceous. As they died back, fast-growing, fast-spreading angiosperms took the opportunity to invade. Studies of fossil pollen from North America revealed the poleward march of flowers during the middle and Late Cretaceous. In one spot in Utah, studies of fossil pollen revealed a forest community that included conifers, ginkgos, ferns, and a few flowering plants. However, sediments from a small pond yielded only angiosperm leaves. The pond sediments were deposited during one of the brief hot spells at the beginning of the Cenomanian age and suggest that flowering plants sought out ephemeral habitats like this during their scramble to the poles.

Warm Cretaceous climate also contributed to an increase in wildfire frequency, which nudged the evolution of remarkable disturbance tolerance in flowering plants. Wildfire also provided some of the oldest flower fossils yet discovered: minute flowers preserved in charcoal from Valanginian- and Barremian-age sediments of Portugal. The fossils captured the delicate structure of these ancient flowers in three-dimensional detail and revealed the surprising diversity already present by that time. Moreover, their preservation in charcoal showed that early flowering plants thrived in fire-prone environments.

A GUIDED WALK THROUGH THE MESOZOIC FORESTS

Time travel is the stuff of science fiction, but geology offers clues to past environments. Coupled with an understanding of living plants and ecosystems—and a little imagination—the fossil record allows us to take a stroll through ancient landscapes.

TRIASSIC AND JURASSIC

Your geological time ship lands first in the midlatitude lowlands of Pangaea, about 252 million years ago. Craggy Appalachian peaks to the south are shrouded in cloud. You open the door to a silent world. The end-Permian extinction event is over, but the insect world was so decimated that the only sound is the wind rustling the leaves of the *Pleuromeia* lycophytes that extend in dense stands as far as you can see. Their stout, unbranched stems grow straight from a beanbag-like base. These must be *Pleuromeia sternbergii*, as the tallest are about 2 m (6.5 ft) high. Their stems are covered in thin, straight leaves that give the impression of giant green bottle brushes. A single, cone-like structure about the size of your fist perches atop some of the tallest plants. The air is hot, heavy, and humid, and the narrow *Pleuromeia* leaves provide little shade. There are virtually no other types of plants here. A few fern clumps grow in a sheltered nook beside an enormous, rotting log. Some tall, dead trees stand amid the stubby *Pleuromeia*

forest. One, which sports the flat, shelflike fungus, is surrounded by a ring of slender shoots. The fan-shaped fringed leaves reveal that it is a ginkgo of some sort. A closer look reveals that the leaves are attached directly to the stem rather than on a short stalk as in living *Ginkgo biloba*. This tree might be *Sphenobaiera*, one of the Late Permian ginkgos that survived the runaway warming at the end of the period. Gingkos, after all, retain buds at the base of their trunk, ready to sprout if the main trunk dies.

Motion catches your eye. Turning quickly, you see a pair of lizard-like creatures scatter from a hollow at the base of another of the dead trees. Behind them, *Moschowhaitsia*, front paws and snout covered in black mud, emerges from the labyrinth of decomposing logs. This four-footed carnivore has a stout body and a head longer than your hand. It must be a juvenile because you recall that the animal's Permian-age ancestors grew much larger. You notice the *Moschowhaitsia*'s sparse coat of spiky fur and recall that it is warm-blooded. Its meal having escaped, the *Moschowhaitsia* locks eyes with you briefly, bears its menacing yellow fangs, and clambers over the downed trees in search of other prey. You latch the door of your time ship and set the date to 167 million years ago to survey the world fully recovered from the extinction.

Landing in the Middle Jurassic, you step out into a coastal forest just as the sun is about to slip below

Spiky tree-sized lycophytes called *Pleuromeia* were among the first large plants to recolonize marshy areas after the end-Permian environmental collapse. These small trees were descendants of the giants of the Paleophytic and persisted for only the first few million years of the Triassic period.

By the Jurassic, cool, moist climates allowed conifers, ginkgos, cycads, and ferns to reign. These plants presented tough, resinous, and toxic forage for the large herbivores of the time.

the western horizon over the young Atlantic Ocean. This part of Pangaea has been drifting apart for about 70 million years. Your time ship has landed amid the tall, straight trunks of giant conifers. A deep-green carpet of moss coats their rough bark. You crane your neck and squint into the upper branches, hoping to identify the trees. Perhaps they are *Araucarites*, their leaves spoon-shaped and with spiked tips that warn they are tough and sharp. Right, you think. This is the time of long-necked sauropods that can browse the uppermost branches. You also notice the scaly, gray-green leaves of *Brachyphyllum* high in the canopy, but the conifers grow so closely together it is hard to tell which leaves belong to which trunk.

Stepping onto the soft carpet of decomposing leaves beneath your feet, you breathe the spicy-sweet smell of conifer pitch. The air is cooler than during the Early Triassic and buzzes with the twilight song of crickets. At your feet you notice a cluster of needlelike leaves attached to a sturdy branch. They look a little like pine needles, but softer, brighter green, and more flexible. Picking up the branch, you notice a string of delicate, scallop-shaped structures about the size of a coin dangling amid the fringe of leaves. This is not a pine but *Czekanowskia*, a relative of the ginkgos, and these are its seeds. A closer look at the tall trunks reveals subtle differences in their bark. There are so many kinds of trees here but they all have the tall, straight trunks. As the sun dips toward the horizon, the understory grows dark. You set off toward a patch of lingering light in the distance, threading carefully through the tufts of deep-green cycad leaves nestled in the spaces between the conifers. Running your fingers across the soft, fur-like covering of the cycad seed cones, you notice they are swarming with beetles. Probably pollinating. Wait! Some of the cycad-like trunks display stout, flowerlike structures. A ring of sturdy, pale pink, leaflike scales surrounds a central sphere where the egg-bearing structures hide among a protective armor of woody barbs. This is a member of the Bennettitales, one of the many Jurassic seed plants that is experimenting with flower-like reproductive structures tailored by natural selection to attract insect pollinators.

As you approach the edge of the forest, ferns become more numerous. You wade into the waist-deep fern foliage at the forest's edge and emerge at a slow-flowing river, its sandy banks covered with a thick stand of horsetails. Crashing through the ferns to the river's edge, you surprise a *Stegosaurus*, drinking just a few meters upstream. It mistakes you for a carnivorous *Proceratosaurus* and raises its spiked tail in defense. It seems best to stay out of sight until the *Stegosaurus* lumbers away. As your heart slows, you notice a surprising diversity of ferns: *Angiopteris*, *Aspidistes*, and *Todites* growing in lush, tangled clumps. *Dicksonia* tree ferns line the edge of the forest and tilt their large fronds toward the open sky like satellite dishes. A shrub emerging from the fern scrum is the one you were looking for—*Caytonia*. This enigmatic plant also evolved features similar to those of flowering plants. Its compound leaves have a slender petiole from which four tongue-shaped leaflets extend like the fingers of a chubby hand. The veins of the leaves intertwine in a complex network that can efficiently supply water across the photosynthetic surface. This one has seeds dangling like tinsel from the base of leaves on the lower branches. Each fertile strand has seven to ten pairs of globular balls on short stalks, with each ball formed from a flap of tissue wrapped tightly around four egg-bearing ovules. Pollen, or perhaps a pollinating fly, could slip into the tiny gap in the enrolled flap, but the nutritious ovules themselves are hidden from insect predators. This seems like a good evolutionary investment, because some of the older leaves are riddled with holes made by hungry herbivorous arthropods. Toward the top of the shrub, similar strands sport pairs of tubular pollen organs that dust your fingers with bright yellow grains. With your magnifying hand lens, you note that the pollen looks very conifer-like.

As the last light fades, you return to your time ship and set course for the Early Cretaceous.

EARLY CRETACEOUS

Your time ship lands next on the eastern shores of North America, 125 million years ago. The ancient Appalachian Mountains still dominate the landscape. Great rivers capture the rain that falls on their highlands and meander across coastal plains to the sea. Although puffy clouds drift overhead, they drop little of their rain on the coastal plain. You landed on a bluff above the river. Here, conifers bearing *Hirmeriella* foliage grow in patchy stands. The growing tips of these conifers have small, scalelike leaves that press tightly against branching twigs. Tender bits at the tips form a spiky tuft at the end of each branch. *Hirmeriella* makes a sparse canopy, and dappled light bathes the forest floor. There, squat cycads and the dried foliage of ferns bake in the sun. You dig the toe of your boot into the soft, sandy soil and kick up a cloud of dust—it is the dry season. The trunk nearest you bears scars of a past fire. Damaged bark has begun to curl around the charred wood beneath in the tree's slow effort to cover the wound. The fire must have been some time ago, because the ground is now covered with a thick layer of dry leaves. The day is hot but not particularly humid. Even so, sweat beads on your nose. As you wipe it away, you jump with a dozen sharp pinches around your ankle. Your boot has disturbed an ant colony, and they are not pleased. You shoulder your pack, stamp your foot free of ants, and head down the hill toward the river below. Scrambling down a north-facing slope, you plunge into a cascade of ferns. A thicket of the drought-tolerant fern *Dryopteris* rims the gully. Descending farther into the cool shade, tree ferns bearing *Thyrsopteris* and *Cladophlebis* foliage reach for the sky. Beneath them, *Aspidium*, *Gleichenia*, and *Anemia* ferns cascade down the slope. On the flat floodplain below, tall, straight *Sequoia* trunks tower above the cypress and yew that have colonized wetter hollows in the landscape. Still more ferns dot the forest floor, and flies buzz annoyingly around your head.

The fern thicket grows dense around the edge of a pond formed where a meandering branch of the river separated from the main channel. The water is still, riffled only by fish breaking the surface to grab an insect. A turtle basks on a nearby log. Across the pond, the beady eyes of a submerged crocodile peer at you over a floating water lily. It is the first flowering plant you have seen so far. You pick your way slowly along the edge of the pond, aware that another croc might be waiting in ambush. You are about the size of one of the croc's favorite prey, the midsized dinosaur *Archaeornithomimus*.

You have heard that the carnivorous theropod dinosaur *Acrocanthosaurus* also prowls these forests, but the dense cypress thickets are probably difficult for the 10 m (33 ft) long predator. The much smaller and even more ferocious *Deinonychus* might be another matter, so you pause between steps to listen for movement in the vegetation.

Soaked in sweat, you finally emerge at the river's main channel. Scanning the sandy bank, you find both shores lined with angiosperms. Herbs, vines, and small shrubs of a dozen species festoon the riverbank, basking like the crocodiles in the bright sun. Their leaves are broad and net-veined and come in a variety of shapes. The leaves seem flimsy compared to those of the ferns and conifers on the floodplain behind you, and insects have riddled them with holes. Some riverside plants display tiny flowers swarmed by clouds of pollinating thrips. Butterflies drift above the scramble of greenery looking for a nectar treat. Other plants display clusters of fire-engine red fruits to potential dispersers. A rustle beneath the foliage makes you jump. Too small for *Deinonychus*, so you drop to your knees and search. Lifting a large leaf, you spy *Argillomys*, a shrew-sized mammal, sitting frozen in fear with a seedpod clenched in its tiny jaw. With a smile, you gently lower the leaf and hear the mammal scurry away. You have been so focused on discovery you did not notice that the morning's puffy white clouds have turned gray and ominous. Wind rises and cold air from the approaching thunderstorm flows over you. It feels wonderful on the hot afternoon, but you do not want to be caught out. Storms in the hothouse world are violent. You feel the hairs on the back of your neck rise, and before you can think to take cover, a bolt of lightning strikes one of the conifers at the top of the bluff. Flammable resin explodes, and the dry debris quickly catches fire. A line of low flames and thick smoke spread westward, driven by the storm winds. Time to get going.

LATE CRETACEOUS

The next stop on your time ship is the eastern shore of North America's western landmass, Laramidia. You check your chronometer before heading out—66

Warming and drying climate, coupled with frequent disturbances like wildfire, allowed flowering plants to spread out of the understory of tropical forests and begin to migrate toward the poles and into a wide range of habitats. During the early Cretaceous, angiosperms were mostly still confined to streamsides and other unstable habitats.

million years ago. Glancing up, you notice that the midday sun is dusky red. You remember that the Deccan volcanoes on the other side of the globe have been erupting for more than 100,000 years now. Moreover, a line of explosive volcanoes dot the western edge of Laramidia. Who knows when one of those last spewed ash into the stratosphere?

To the east, at the edge of your vision, you see a ribbon of white sand and the glimmering blue of the Cretaceous Western Interior Seaway. The sea extends from the Gulf of Mexico to the Arctic, dividing North American into two landmasses. Turning to the west, you see the low ridges of the Rocky Mountains and below you to the north a wide, brown river winding slowly toward the sea. The river is full of sediment washing down from the rapidly eroding Rockies. A flock of pigeon-sized *Ichthyornis* rest on a sandbar. These fish-eating birds are usually found much closer to the sea, but they may have followed the spawning gar upstream. A few meters upstream from the resting birds, three large *Edmontosaurus* duck-billed dinosaurs tug at a patch of water lilies growing in a quiet bend of the river. A fourth, much larger *Edmontosaurus* stands guard on the levee above the resting birds.

The landscape is a patchwork of greens. Swatches of deep-green cypress and *Glyptostrobus* dot the low, marshy parts of the landscape. On drier parts of the river's floodplain, a low woodland of emerald and apple green drapes over the rolling hills. Unlike the straight trunks of Jurassic and Early Cretaceous conifers, most of these trees are gnarled and irregular, with the globular crowns of angiosperms. Many of these trees have large leaves, suggesting that rainfall is plentiful. However, their leaves are relatively thin, a trait common among plants evolved for climates cool enough to encourage trees to shed their foliage periodically. Beyond winter, it makes good evolutionary sense for the trees to refresh their foliage periodically, because butterflies flutter everywhere above the woodland canopy, and their hungry larvae lurk below. The irregular carpet of trees is studded here and there with the tall, straight trunk of the dawn redwood (*Metasequoia occidentalis*) and an occasional *Ginkgo*. These elders of the Mesozoic have now become rare. Amid the woodland are other gray-green patches of fern thicket where the trees have been temporarily knocked back by some disturbance. The patches of bare brown earth amid the ferns provide a clue. Perhaps these are wallows for herds of *Triceratops* that munch their way through the coastal woodlands. There are no *Triceratops* in sight today and that is good. The large predators that feast on these three-horned herbivores are probably following the herds. It might be safe to descend into the forest.

Where the Jurassic forests were an orderly cathedral of straight trunks, the Late Cretaceous forest is a tangle of irregular branches. Trees sporting compound leaves with slender, serrated *Dryophyllum* leaflets and heart-shaped *Celastrus* leaves are common. Spiky palms scrabble through the understory, and fruit drips from many branches. The forest floor is also different, here covered in a thick layer of decomposing leaves. Tree ferns stretch for the canopy, too, extending their long petioles to compete for light with the fast-growing angiosperm trees. A rare *Nilssoniocladus* cycad shelters in their shade. In a bit of irony that makes you chuckle, an inconspicuous tuft of green nestles at the roots of the cycad. It is one of the early grasses that will—millions of years from now—diversify and dominate this place. In the latest Cretaceous, though, they are newcomers. Scrabbling through the undergrowth, you reach the edge of one of the fern thickets and push through the dense tangle of *Vitis* vines that line the clearing. As you scan the clearing, a smear of light in the southern sky catches your attention. It's too low to be the moon and you remember: Today is the day the Cretaceous—and the Mesozoic—will end in a fiery burst of white-hot rock—impact day. You look around for one last moment. The *Edmontosaurus*, *Triceratops*, and *Tyrannosaurus* will all be gone soon. The trees will be stripped of their foliage by the force of the blast and many killed. But this vegetation is accustomed to disturbance, even cataclysmic disturbance. And the Earth will green again soon. You, however, are too big to survive the catastrophe. No animals larger than 25 kg (55 lb) remained after the impact. You return to your time ship and leave the Mesozoic.

By the late Cretaceous, flowering plants had evolved a wide range of sizes including woody trees and shrubs, and small herbs. They also formed the backbone of most plant communities. Conifers and ginkgos still thrived, but in isolated stands or specialized habitats. And a new group of ferns, the Polypodiales, diversified in the deep shade of the new flowering forests.

IF FLOWERING PLANTS HAD NOT BECOME DOMINANT...

Looking at the history of life on Earth from the vantage point of the present day, the evolution of the organisms and ecosystems we see around us seems inevitable. The reptiles and dinosaurs of the Mesozoic must have been inferior to the mammals that fill the same ecological roles today. Despite their 165-million-year run as the most important large animals on land, "dinosaur" remains the term for something that is not up to the rigors of the modern world. However, from a Mesozoic perspective, it would be hard to imagine that the tiny, ratlike creatures scuttling in the underbrush would someday evolve into lions, bats, wolves, antelope, giraffes, apes, elephants, and whales—the tremendous variety of modern mammals. The happenstance of the end-Cretaceous extinction removed dinosaurs and opened the evolutionary door to mammals. Catastrophic extinction events like those that started and ended the Mesozoic are unpredictable and reset the evolutionary stage, opening opportunities for new lineages at unexpected moments. Or so it goes with animals. The Mesozoic plant story is different.

Many species of Paleozoic plants perished during the end-Permian extinction, but most major groups survived and diversified again in the Triassic. Another extinction event at the end of the Triassic wiped out many more plant species, but again, most major groups persisted to generate still more diversity in the Jurassic. Again, at the end of the Cretaceous, a major extinction event wiped out a host of plant species, but the fundamental nature of Late Cretaceous plant communities changed little as the Cenozoic began. Instead, the Mesozoic botanical transformation happened over tens of millions of years between the second and third animal extinctions. A new lineage born in the steamy understory of the Mesozoic tropical forests, the flowering plants, evolved a suite of features that together allowed them to outcompete Triassic and Jurassic dominants, particularly when it came to exploiting disturbed sites. Given that many Jurassic seed plants independently evolved features like those key to angiosperm success, it would be easy to imagine that some upstart plant would eventually put all the features together and begin the revolution. But that perspective comes from looking backward. During the Mesozoic, there was no such guarantee. Without the lucky combination of pollinator relationships, high photosynthetic rates, and rapid growth and reproduction, the remarkable diversity and ecological dominance of flowering plants would not have been possible. What might the world be like without them?

FLOWERING PLANTS AS RESOURCES FOR ANIMALS

Flowering plants pack their seeds with high-energy deliciousness (carbohydrates, fats, and protein) to sustain their embryos through long periods of seed dormancy and during the early days of growth in the understory where the photosynthetic capacity of the youngest leaves might not be enough to support rapid growth. Many angiosperms also wrap their seeds in sweet and nutritious fruit to attract and reward seed dispersers. In addition, the super-charged photosynthetic system of angiosperms renders their leaves higher in protein than the foliage of other plants. And many flowering plants store their photosynthetic harvest in their roots as starch, which provides another animal food source. Flowering plants provide more calories and more nutrition bite for bite than the plant groups that preceded them.

For Triassic and Jurassic herbivores, the hunt for food was about processing ever larger quantities of low-quality forage. In dinosaurs, groups like the sauropods evolved large size to become gigantic fermentation tanks where microbes broke down indigestible cellulose. As flowering plants became more numerous in the mid-Cretaceous, many herbivorous dinosaurs shifted toward forms that foraged closer to the ground, where most of the early flowering plants grew and fruit fell. Without flowering plants, the great herds of Cretaceous horned herbivores could probably not have found enough to eat. Energy-hungry mammals also took advantage of the plentiful resources offered by angiosperms. Mammals needed the abundant high-energy food provided by flowering plants to fuel their own evolutionary diversification. While the extinction of dinosaurs may have been necessary to open ecological opportunities for mammal evolution, the extinction alone was not sufficient. Without the abundant calories provided by angiosperms, we might not be here. Without flowering plants, the end-Cretaceous extinction might have ushered in the Age of Turtles and Crocodiles, rather than the Age of Mammals.

MESOZOIC PLANT CONSERVATION

Most of the plants we eat and the plant-derived medicines we use come from flowering plants. When asked which non-angiosperms humans regularly consume, most people in Western cultures struggle to name even a few. Anthropologists estimate that humans have domesticated about 200 plant species for food, virtually all of them flowering plants, and 90% of our plant-based calories come from just 30 angiosperm species.

Indigenous and traditional cultures also see the botanical world as a pharmacy, with angiosperms making up most of the healing diversity. Western medicine has co-opted some of these plants, learned their biochemical secrets, and formulated commercial versions of these medicines. Among the most common examples of angiosperm-derived medicines are aspirin, from willow (*Salix*) bark, natural opioid medicines like morphine and codeine, prepared from the poppy species *Papaver somniferum*, and the heart medication digitalis, extracted from foxglove (*Digitalis purpurea*). Quinine, from the bark of the cinchona (*Quina*) tree, was used in the nineteenth and early twentieth centuries to treat and prevent malaria, and in the late twentieth century, artemisinin, produced from the sweet wormwood (*Artemisia annua*), became the standard of care for the disease. Vincristine and vinblastine, made from Madagascar periwinkle (*Catharanthus roseus*), revolutionized chemotherapy for lymphoblastic leukemia, non-Hodgkin lymphoma, and testicular, ovarian, breast, bladder, and some lung cancers in the 1960s. Their introduction increased the survival rate of several childhood cancers from 10% to nearly 95%. Despite the dominance of flowering plants in the human pantry and medicine cabinet, some Mesozoic survivors hold important economic and cultural significance and continue to thrive through partnerships with people.

TAXUS: A TREATMENT FOR CANCER

One conifer from an Early Cretaceous lineage provides a potent anticancer drug. The Pacific yew (*Taxus brevifolia*) produces paclitaxel (known by the trade name Taxol), used to treat and prevent recurrence of several types of breast, ovarian, esophageal, cervical, pancreatic, and lung cancer. Paclitaxel was first isolated in 1962, and its anticancer properties were described in 1971. The U.S. Food and Drug Administration approved the first commercial formula for clinical use in 1993. Some considered it the most important chemotherapeutic drug introduced in the last 50 years. When Taxol first came to market, it could only be isolated from the bark of wild Pacific yew. Because harvesting bark destroys the growing cells beneath, paclitaxel production killed the trees. Yew populations quickly came under significant pressure, and because they grow slowly, trees could not be harvested sustainably to meet the tremendous demand for the life-saving drug. A single cancer patient, for example, required the bark of approximately eight 60-year-old yew trees for their initial therapy. Within a few years, many wild populations were threatened.

Taxus brevifolia grows in the forest understory and along exposed coastal hillsides in the Pacific Northwest of North America, along the coast from northern California to southernmost Alaska and following river valleys inland into southeast British Columbia and northern Idaho. Preferring moist soil, it tends to establish along streamsides or in areas with abundant precipitation, like the coastal rainforests of Oregon, Washington, and British Columbia. Since the US FDA approved paclitaxel, an estimated 30% of the wild populations have been depleted. Moreover, several decades of climate change have produced hotter and drier summers across the tree's home range and increased the frequency of wildfire. *Taxus brevifolia*'s thin bark and ground-hugging growth habit make it particularly vulnerable to fire, and its seeds seldom survive to germinate after adult trees burn. Today, the tree is threatened across much of its range, prompting efforts to find other ways to produce the life-saving medicine.

Drug manufacturers have taken several approaches to providing a sustainable supply of paclitaxel. In 1994, researchers succeeded in synthesizing paclitaxel in the laboratory. However, the process was complicated, and yields remained low, rendering the process uneconomical. The trees themselves remained the best source. Next, researchers explored the other 11 living *Taxus* species for new sources of paclitaxel. However, none could match the potency of *Taxus brevifolia*. Scientists also experimented with harvesting paclitaxel from yew leaves rather than bark and, in the process, discovered a fungus that had acquired the ability to make paclitaxel. However, the fungus proved difficult to culture in the lab and yielded only a little paclitaxel. In Europe, competing research teams developed partial synthesis strategies, starting with a natural plant chemical derived from *Taxus* leaves and finishing the synthesis in the lab. Using natural compounds derived from European yew (*Taxus baccata*), several methods emerged. And because the starting products are abundant in yew leaves, commercial-scale manufacture required less plant material than that needed to extract paclitaxel directly. Today, most paclitaxel is manufactured with the partial synthesis method, using yew leaves or the same natural chemicals derived from laboratory cell cultures. Both methods significantly reduce the pressure on wild

The bark of *Taxus brevifolia*, the Pacific yew, contains a powerful cancer-fighting chemical.

populations of yew and allow trees to slowly recover. Human ingenuity is allowing a plant that has given so many the gift of life an opportunity to thrive too.

CYCADS: ANCIENT CULTURAL CONNECTION AND MODERN EXPLOITATION

Cycads were so abundant and diverse that the Triassic and Jurassic are called the "Age of Cycads." However, slow growth and reproduction doomed them during the Flowering Plant Revolution. Today, about 300 species remain. Many of the species that maintain healthy populations in the wild have found partnerships with humans. Genetic and ethnobotanical evidence suggest that humans have tended wild cycads as food and cultural resources since humans first inhabited the Ryukyu Islands of Japan more than 35,000 years ago. In the South Pacific nation of Vanuatu, people traditionally plant and tend cycads to mark important historic and ritual sites, and cycad leaves symbolize reconciliation and authority. Indigenous cultures in Central Mexico use the seeds of several species of *Zamia*, *Dioon*, and *Ceratozamia* in shamanic rituals, where participants consume preparations made from the toxic seeds to enter altered states of consciousness. Seeds and pollen of *Encephalartos* are used in a similar way in the Limpopo region of South Africa, where illegal harvest for cultural and recreational use puts some populations in peril.

For *Macrozamia communis*, known in the Dharug language as *burrawang*, a partnership with the Aboriginal peoples of southeast Australia allows the plant to thrive in the face of development and environmental change. When prepared in a way that removes toxic cycasin, the seeds of *Macrozamia communis* provide a starchy flour that traditionally provided a seasonal food for Aboriginal peoples and today provides an important cultural touchstone. Like many cycads, *Macrozamia communis* tends to synchronize production of male and female cones to encourage crossbreeding among individuals. Aboriginal peoples understand that they can stimulate this mass production of seeds using cultural burning, the practice of lighting low-intensity fires on the landscape to manage vegetation. By carefully timing burning, Aboriginal peoples produce an abundant crop of seeds and allow seedlings to germinate and grow large enough to withstand the next cycle of fire. Cultural burning also manages flammable debris on the ground to reduce the risk of intense wildfire that would kill the ancient plants. Tens of thousands of years of this management has allowed *Macrozamia communis* and all the other plants and animals in the region to thrive.

However, not all cycads have attentive human caretakers and many are critically endangered or functionally extinct. For example, *Encephalartos woodii* can no longer be found in the wild. The species was discovered by John Medley Wood in 1895 on a steep, south-facing slope in oNgoye Forest in KwaZulu-Natal, South Africa. Wood described four large stems that probably represented growth from a single, much older, root system. In 1899, botanists collected several shoots from

these wild stems for Kew Gardens in the United Kingdom, and in 1903, more shoots were transplanted to the Durban Botanic Gardens, where Wood was curator. After several instances of poaching, the last surviving stem was transplanted to Pretoria in 1916; it died in 1964. Today, the species grows in botanical gardens around the world, but all these plants originated from the single wild individual—a male. No seed-producing plants have been found. While clones from the wild plant may live for centuries in captivity, the species will

The cycad *Encephalartos woodii* is among the rarest of living plants. Only a single pollen-producing individual has ever been found in the wild, making the species functionally extinct.

Pollen-producing (left) and seed-producing (right) individuals of *Macrocycas calocoma*. The species is found only in isolated patches of undisturbed forest in Cuba. It is threatened by development, poaching, and a very low reproductive rate.

never again set seed and is therefore functionally extinct. To rescue the species, botanists used *Encephalartos woodii* pollen to produce hybrid seed with a closely related species, *Encephalartos natalensis*. In this plan, female hybrids will then be fertilized with *Encephalartos woodii* pollen to produce another generation of seeds that will contain a higher proportion of genes from *Encephalartos woodii*. Over successive crosses, the resulting plants will have progressively more and more traits from *Encephalartos woodii*. However, botanists discovered that offspring of these crosses always inherited important cellular structures—chloroplasts and mitochondria—from *Encephalartos natalensis*. Therefore, the strategy can never recreate the genetic distinctiveness of seed-producing *Encephalartos woodii*. For now, extinction is still forever.

Many other cycad species, including *Encephalartos latifrons* in the Eastern Cape Province of South Africa, are critically endangered by habitat loss and poaching for the horticultural trade. Although it is native to an area conserved for its botanical diversity, only about 70 individuals are left in the wild and most are lone plants separated from others by a kilometer or more. No natural reproduction had been observed for more than 50 years. In 2013, researchers discovered a new population of about 30 plants, nearly a third of which were seedlings. Throughout most of its range, male *Encephalartos latifrons* outnumber females by about four to one, but in the newly discovered group, there are approximately equal numbers of male and female individuals, suggesting that older plants are reproducing successfully and the population may be capable of sustaining itself. The location of the new stand remains secret because rare cycads can be valuable. In August 2014, 13 *Encephalartos latifrons* individuals were stolen from the Kirstenbosch Botanical Garden in Cape Town, South Africa. Those plants sold for more than US$20,000 on the illegal market.

GINKGO: A SACRED PLANT IN SOUTHEAST ASIA

Plants that were unmistakably part of the ginkgo lineage first appeared in the Permian and flourished through the Jurassic and Early Cretaceous. Although they never dominated Mesozoic forests, ginkgos were always present and evolved significant diversity. Like the cycads, with which they share a common ancestor, ginkgos began a slow decline around 100 million years ago when flowering plants surged to dominance. By the end of the Mesozoic, only a handful of species remained. Ginkgos survived the end-Cretaceous extinction and maintained a worldwide distribution for another five million years or so until they became extinct in the Southern Hemisphere. *Ginkgo* followed the dawn redwood (*Metasequoia glyptostroboides*) into refuges in Asia as the interiors of continents dried and climate cooled during the Cenozoic. By the time humans arrived in the mountains of present-day central China some 50,000 years ago, *Ginkgo* was struggling to hang on in a handful of isolated populations.

The edible seeds of *Ginkgo* probably first attracted human attention. After the soft outer covering is removed, the hard, nutlike seed can be boiled, roasted, or stir-fried. *Ginkgo*'s unusual features developed deeper cultural significance. Older trees produce downward-growing shoots from their lower branches. When shoots reach the ground, they root and begin to produce new upward-growing shoots, forming a thicket around an ancient parent tree. This habit evoked the protective and nurturing qualities of motherhood, and *Ginkgo* became a totem of protection for new and nursing mothers. Upon their arrival in China around 200 BCE, Buddhist monks recognized *Ginkgo*'s longevity—individuals can live for 3,000 years—and associated it with the wisdom acquired from a long life of treelike contemplation. Monks planted *Ginkgo* around their settlements and carried seedlings when they traveled. Genetic evidence suggests that the ancient *Ginkgo* population on West Tianmu Mountain in Zhejiang Province, once thought to be a relic of the wild population, was brought to the region about 1,500 years ago by traveling monks. Buddhist explorers also introduced the tree to Korea and Japan, where it became an important symbol in the Indigenous Shinto tradition.

The last survivor of the ginkgo lineage, *Ginkgo biloba*, is revered for its longevity and resilience. Mature plants produce prop roots that grow from the lowest branches to the ground and give rise to new shoots.

In Japan, *Ginkgo* first caught the attention of European botanists. Engelbert Kaempfer, a physician with the Dutch East India Company, noted the tree's graceful, fan-shaped foliage and described it in his 1712 book, *Amoenitatum Exoticarum*. Kaempfer coined the word "ginkgo," which was probably a transliteration of a regional pronunciation of the Japanese *kanji* used for the tree. Carl Linnaeus did not mention *Ginkgo* in the first edition of his *Species Plantarum* in 1753, even though a few seedlings had already been planted in Europe by that time. Linnaeus named the species in *Mantissa Plantarum II*, published in 1771, calling it *Ginkgo biloba*, probably because the example he had at hand came from a young shoot in which the familiar fan-shaped leaves are bisected by a deep furrow to create two "lobes."

Today, *Ginkgo biloba* has transcended its status as a Mesozoic relic thanks to humans. Following the Great Kantō Earthquake and firestorm that destroyed Tokyo and Yokohama in 1923, the Japanese government evaluated a variety of trees to landscape the reconstructed cities. They looked for species that would be fire-resistant, long-lived, and hardy in urban pollution. *Ginkgo*'s tolerance for a wide range of climate conditions, polluted air, drought, and shallow, poor-quality soil quickly made the species a popular urban tree worldwide. By some estimates, the five boroughs of New York City host an urban forest of more than 60,000 *Ginkgo* trees. The popularity of *Ginkgo* as a street tree has allowed the species to regain almost all of its worldwide Mesozoic range.

EPHEDRA: A WORLDWIDE MEDICINE

The seed plant genus *Ephedra* first appeared in the Early Cretaceous and is known from the fossil record of North America, Argentina, China, and Portugal, suggesting that it had a worldwide distribution. Today, nearly 70 species are found across all continents except Antarctica. Living *Ephedra* prefers arid climate and high elevation, which may explain its near absence from the fossil record, except as dispersed pollen. *Ephedra* produces a variety of compounds that have widespread medicinal applications both in traditional and westernized medicine. In China, *Ephedra sinica* (known as *mahuang*) has been used to treat asthma for more than 5,000 years. In Ayurvedic medicine, *Ephedra gerardiana* treats allergies, asthma, headache, and nasal congestion. And in North America, various Indigenous cultures used their local *Ephedra* species for a variety of medicinal purposes. Havasupai, Hualapai, Mahuna, Navajo, and Paiute Nations from the southwestern United States brew infusions of various parts of the plant to treat digestive and kidney problems. O'odham people of the Sonoran Desert treated syphilis and gonorrhea by preparing infusions of *Ephedra antisyphilitica* to drink and apply topically. This use earned the species its common colonizer name, "whorehouse tea." Shoshoni people dry, grind, and steep *Ephedra* seeds

Plants in the genus *Ephedra* are used medicinally in cultures worldwide. Species like *Ephedra antisyphilitica* have separate pollen- and seed-producing individuals.

Pollen-producing   Seed-producing   Pollen-producing   Seed-producing into a stimulating, coffee-like beverage. Tewa people chew the leaves and stalks to reduce diarrhea. And in Mexico, the diuretic effects of *Ephedra pedunculata* made it an important traditional treatment for pleurisy, pneumonia, and kidney failure.

Western medicine has also embraced several of *Ephedra*'s medicinal compounds. Chemists isolated 26 biologically active molecules from *Ephedra*, including pseudoephedrine and ephedrine. Pseudoephedrine is a common decongestant. Ephedrine is commonly used as a stimulant to prevent dangerously low blood pressure during anesthesia. Today, most commercial preparations of pseudoephedrine and ephedrine are chemically synthesized so their use does not endanger wild plant populations. Preparations from whole plants have found a market as herbal supplements, but most species are common, if not widespread, so harvesting does not endanger most populations. In fact, *Ephedra*'s hardiness makes it a good choice for rehabilitating degraded land in dry climates, so the range of most species is expanding in this collaboration with humans.

KAURI: A RACE TO SAVE MESOZOIC GIANTS

Agathis australis, also known by its Māori name kauri, is a modern descendant of a lineage that extends to the Middle Jurassic. Today restricted to New Zealand, Australia, and parts of Southeast Asia, *Agathis* once grew in southern South America. Mature kauri are generally the largest, if not tallest, trees in the forest, with trunks growing to diameters well over 5 m (16 ft). While many conifers failed to compete successfully with upstart Cretaceous angiosperms, *Agathis australis* evolved several strategies that allowed it to hold on to dominance in some subtropical forests. First, kauri have a two-tier root system: Thick, so-called peg-roots penetrate deep into the soil to firmly anchor the trees, which can grow to 50 m (over 160 ft) in height, while an extensive network of fine roots, partnered with symbiotic fungi, extends through the organic layer on the soil surface. The fine roots scavenge nitrogen and phosphorus released from decomposing leaf litter before these nutrients can filter into the mineral

soil. This strategy successfully intercepts key nutrients and starves more deeply rooted angiosperms. Second, kauri discourages climbing angiosperms by dropping lower branches and infusing its flaking bark with defensive chemicals. Third, kauri bark continually sloughs off and accumulates in a thick, mulch-like layer that discourages the germination of competing plants. Fourth, kauri bark and foliage are rich in organic acids that, when mixed with rainwater, leach mineral nutrients like iron and magnesium out of the soil's upper layers and deliver them to deeper horizons of the soil. Here, the kauri's deep-penetrating roots harvest the mineral bounty.

Despite this repertoire of evolutionary advantages, kauri in New Zealand are falling victim to a disease that threatens to wipe out the species. The fungus *Phytophthora agathidicida* infects the fine surface root system of the kauri, choking off its ability to gather nutrients and water. Eventually, the fungus spreads into the wood, causing leaves to yellow and trunks to rupture and bleed resin. Nearly all the trees that begin to show these symptoms eventually die. Scientists remain perplexed about the origin of the disease, known as kauri dieback. An initial outbreak was reported in the 1950s, linked to a different species of *Phytophthora* that was believed to be widespread in New Zealand. In 1972, a new outbreak emerged on Great Barrier Island/Aotea about 100 km (62 mi) northeast of Auckland. This time, the disease was linked to *Phytophthora agathidicida*. In 2006, forest stewards discovered infected kauri in the Waitākere Ranges on the North Island mainland west of Auckland. Scientists believe that the disease-causing fungus spread on the boots of hikers and expressed concern about ongoing human access to the region. In response, the local Māori community, Te Kawerau ā Maki, held a ceremony in 2017 that proclaimed the region under sacred protection, which made entering the area taboo. Within a few months, the New Zealand government closed parks across the region. In 2022, the government, in collaboration with local Māori communities, announced Tiakina Kauri, a national plan to save the kauri. By 2024, protection measures had allowed the reopening of some iconic kauri stands, under the watchful eye of Māori guardians. However, the disease was not yet under control.

Agathis australis, known as kauri, grows only in northern New Zealand. Kauri have evolved several strategies to compete with faster-growing flowering plants, which allow kauri to remain important members of the native forests.

WHERE TO FIND MESOZOIC PLANTS

Fossil plants are commonly found in sedimentary rock—rock that began as particles of sand, silt, and clay washed around in the environment. Sedimentary particles are transported by wind, water, or ice, then deposited in a particular place, compacted, and stuck together to become rock. If plant parts become trapped in and deposited with that sediment, they can become fossils.

Although the connection between a leaf impression on a slab of rock and a living plant seems obvious to a modern mind, that understanding is a relatively recent discovery. Ancient Greek and Roman philosophers collected and considered fossils. They acknowledged their similarity to the living creatures but classified them as "rocks." As late as the seventeenth century, thinkers could not make the intellectual leap from fossil to living organism because they could not imagine how nature placed a living creature *within* rock. When Danish clergyman Nicolas Steno connected sand carried by a river with the layers of sedimentary rock visible on land, the notion that an organism could be buried

and transformed to rock along with the sand slowly emerged. Short decades later when Kaspar Maria von Sternberg and Adolphe-Théodore Brongniart began splitting open layers of shale, the idea that these leaves, stems, and cones once grew on living plants was well established. Early geologists had also connected the sedimentary rocks themselves to a diversity of environments where such rocks could form. These insights not only gave context to the fossils themselves but also allowed paleobotanists to reconstruct the physical and biological environment in which fossil plants once grew.

HOW PLANTS BECOME FOSSILS

One of the most vexing truths about the plant fossil record is that most plants fall apart as they begin their journey from living organism to fossil. Leaves fall from trees, pollen and spores are cast far from the

Plant parts like leaves fall from the parent plant. If they land where sediment accumulates, like a lake or river, they may sink to the bottom and be buried by sediment and preserved.

parent plant, branches break, stems are severed from roots, flowers and cones fall, seeds disperse, and fruit is nabbed by an animal and carried off. Finding a whole plant—root to shoot, complete with leaves and reproductive structures—is so rare that such fossils become instant celebrities. One example is *Archaefructus liaoningensis*, an aquatic herb from the Yixian Formation in northeastern China. The clay-rich sediment in which the fossil was preserved hinted at a lake environment, with Cretaceous-age *Archaefructus* rooted water lily–like in the shallows. A handful of plants were somehow uprooted and drifted into deeper water, where they quickly sank and were buried by subsequent layers of clay-rich sediment. *Archaefructus* happened to be in bloom when it was disturbed, which told its discoverers that the plant was among the earliest angiosperms. Initially, in fact, they described the fossil as Late Jurassic in age because the oldest rocks from the Yixian Formation were indeed Jurassic. However, subsequent detailed age dating of the sediments surrounding the fossil itself confirmed that it was 125 million years old—Early Cretaceous in age. However, even though *Archaefructus* was dethroned as the oldest angiosperm, it remains a vitally important fossil because we see the plant preserved much as we might have found it in life, a rare gift in the world of fossil plants.

To understand the journey of a more mundane fossil, imagine a *Ginkgo* growing on the bank of a Cretaceous river on the eastern shore of Laramidia in what is today the state of Montana. At the end of the dry season, a few leaves have turned golden. A gust of wind sweeps one from the tree and it swirls away from its parent, eventually settling on the surface of the slow-moving river. The floating leaf makes its way toward the mouth of the river and gradually becomes waterlogged. As water penetrates the leaf, a myriad of microbes colonizes its surface and begins to feast. The microbial cargo makes the leaf just a bit heavier and it dips below the surface. At just the point where the incoming tide balances the gentle flow, the leaf settles to the riverbed along with clay particles also carried suspended in the stream. The clay coats the leaf, filling in even the tiniest crevices, and smothers the hungry microbes. As the tide slacks, more clay falls, and a centimeter-thick layer accumulates with the leaf tucked inside. The tide turns and once again river flow drives a tumble of sand along its bed, burying clay layer and leaf in even more sediment.

Millions of years later, fossil-containing rocks may erode to reveal their treasure. When recovered with key geological context, fossils can yield clues about ancient plants and their world. Eventually, the fossil should find a home in a museum where it is available for study and exhibit.

Time rolls by. Rain falling on the mountains to the west erodes more sediment that the river carries slowly toward the sea.

Meanwhile, back in the clay layer, a tiny electrical charge in the clay grains binds loosely with the surface of the leaf, molding every detail of the surface. The raised line of the leaf veins and the ragged hole where a long-ago caterpillar munched are all perfectly cast in clay. As more sediment piles up, water squeezes out of the sediment. Most of the organic substance of the leaf also dissolves and washes out of the clay layer. More pressure from the rock above packs the clay tightly, and minerals in the departing water glue the grains together, transforming it to rock. Millions of years pass. Dinosaurs fall; mammals flourish. The continent rises, climate dries, and grasses evolve and spread over the landscape. The little rain that falls on the eastern side of the Rocky Mountains collects into a meandering river that carves deeply into the landscape as the plains rise beneath it. The Cretaceous-age rocks see the sunshine once again.

In Africa, an upstart ape combines a curious brain and clever fingers into technological civilization that spreads to dominate the globe like the flowers of the Cretaceous. Seventy million years later, two of those humans discover some bone eroding from the sediment. In the blink of a geological eye, the paleontology team obtains special permits to excavate the dinosaur whose leg bone they discovered. As the summer begins, a troop of humans with picks and shovels slowly strip away layer after layer of sandstone and clay to reach the trapped dinosaur. One notices a leaf in the pile of castaway rock. Call the paleobotanists!

The paleobotany team in Montana begins by carefully measuring and documenting the layers of rock from which the leaf fossils have come. Then they spend days in the blistering sun meticulously picking through the pile of excavated rock, locating the centimeter-thick clay layers and splitting open each to check for a leaf within. Most are barren, but occasionally a cry of excitement heralds a discovery. Each precious fossil is swaddled in a thick, protective layer of toilet paper, then paper towel, and secured with tape. The date and location are written across the tape and the precious parcels are set gently aside. At the end of the day, the paleobotany team empties their packs of the water they brought to their colleagues and fills them with the carefully wrapped fossils to begin the two-hour trek back to the vehicles and another hour's drive to camp. They will return to continue tomorrow. There are still tons of rock to split.

In camp, the fossils are unloaded, encased in bubble wrap, and boxed for the trip back to the lab. There, the paleobotany team will unwrap each fossil, delicately clean and trim it, and paint a small white patch on an inconspicuous corner of the rock. On the patch, they print two numbers in minute black script: One represents the location and matches a computer record that describes the place from which the fossil came. All the fossils from this site share this locality number. The second number is unique to the specimen and links to a name and description that researchers will eventually give to the leaf. Only then can the fossil find its home in a museum that cares for many such fossils. Together with all the information about the rock and location where it was found, the fossil is now ready for study, which may include description and naming, analysis of the climate under which it grew, the diversity and structure of its forest, and the creatures that gnawed on it, to name just a few of the secrets it might eventually reveal. Although the journey of every plant fossil from its parent plant to the museum is unique, most plant fossils follow a similar path. For some plants, mineral-rich water fills their cells, petrifying them in place; others may be encased in sediment that also preserves their organic structure. Whatever the unique circumstance, all fossil plants must be buried in sediments that later turn to rock, uplifted to the surface where they can be discovered, then collected and preserved for study and wonder.

To find fossil plants, you first need to find sedimentary rocks of the right age and representing the right environment. Because most plants grow on land, marine sediments are not the best place to find plant fossils. However, rafts of vegetation and large logs sometimes float out to sea and can be preserved there, but rocks formed in environments on land offer a richer record. Because of the almost magical properties of clay to slow decomposition and preserve minute detail, rocks formed in muddy environments like lakes and swamps are good places to look. Plant fossils are also preserved in the floodplains of rivers, particularly when buried as sediment-laden floodwaters spill over a river's banks. In some unusual cases, plants smothered in mineral-rich hot springs or buried by volcanic ash can also be preserved. In fact, most terrestrial sedimentary rock contains some plant fossils, even if they are only scraps or microscopic spores and pollen. Whole leaves or cones or large chunks of wood are rarer and tend to be concentrated in relatively small areas. It takes many hours of exploring and splitting open rocks to find a good spot, but once found, such places may yield fossils for many years.

Paleobotanists begin by studying local geology to find sedimentary rocks that are of the right age and represent an environment where plants were likely to have grown and been preserved. Reading the scientific literature and guidebooks helps because previous generations of geologists have already documented many productive spots. Once a collecting site has been identified, research the laws that govern fossil exploration and collecting. Many states and countries have laws that limit collecting or require collectors to hold permits, which can take months to obtain. Therefore, begin to plan well ahead of the time you intend to do fieldwork. Finally, before you head out into the field, be sure to check land ownership and any special rules that apply to the place you intend to explore. For example, exploration and collecting may be entirely forbidden in protected areas. And never trespass on private land, even to reach a fossil site to which you have legal access. Taking the time to work constructively with landowners and other stakeholders is vital to the science of paleontology. And when you head out into the field, don't forget a large supply of toilet paper, not only out of necessity but also because precious plant fossils are unlikely to survive the trip home without careful wrapping.

TRIASSIC PLANT FOSSIL HOT SPOTS

Triassic-age plants can be found worldwide, but a few spots have yielded particularly important insight into the Triassic world.

Pleuromeia lycophytes were among the first large plants to recolonize the scorched landscape after the end-Permian extinction event. *Pleuromeia sternbergii* was discovered in the Buntsandstein sandstone near Mägdeburg, Germany. The species was later found in similar rocks across central Europe, including France, Austria, and Spain. Other species have been collected from Siberia (*Pleuromeia olenekensis*), eastern and central Russia (*P. rossica*), and Kazakhstan (*P. jokunzhica*). The worldwide distribution of this pioneering plant was confirmed with species from Japan (*P. hataii*), northern China (*P. pateriformis* and *P. altinis*), and southern China (*P. sanxiaensis* and *P. marginulata*). Although some sources mention a possible *Pleuromeia* from Australia, the assignment of those fossils to this genus has been questioned.

In North America, the Chinle Formation includes sandstones and shales from the Late Triassic. It represents a region of marshes, swamps, rivers, floodplains, and coastal mudflats that extended over a wide region of what are today's states of Arizona, New Mexico, and Utah. The Chinle Formation is most famous for the large, fossilized logs of *Araucarioxylon arizonicum* preserved at Petrified Forest National Park in Arizona. *Araucarioxylon* was a forest-dominant conifer with a diversity of other plants growing in its dappled shade. At least nine types of fossil wood have been described from the Chinle Formation along with more than 200 other plants, including ferns, cycads, lycophytes, horsetails, ginkgos, and so-called "seed ferns"—extinct seed plants with fernlike foliage.

The Triassic flora of the Southern Hemisphere was dominated by the seed fern *Dicroidium*. The group first appeared in the Late Permian of Jordan and expanded southward following the end-Permian extinction event. As Pangaea broke apart in the Early Triassic, *Dicroidium* became extinct in the Northern Hemisphere and spread across Gondwana to become an icon of the supercontinent's Mesozoic flora until it was supplanted by angiosperms in the Cretaceous. Important Triassic *Dicroidium* hot spots in Australia include the Ipswich Coal Measures west of Brisbane, the Triassic cliffs north of Sydney, and coal mines across South Australia. In New Zealand, *Dicroidium* can be found in stream banks in northern Otago. Still more diversity emerges from the Molteno Formation

Today, the region surrounding Petrified Forest National Park in Arizona, USA, is a desert. During the Triassic, it was a lush forest where fierce storms toppled trees and buried them in sand from flood-swollen rivers.

in the Karoo Basin in southern South Africa, the Ischigualasto Formation north of Mendoza, Argentina, the Rewa and Damodar Basins of northeastern India, Victoria Land in East Antarctica, and Scoresby Sound in East Greenland.

JURASSIC PLANT FOSSIL HOT SPOTS

The most important and comprehensive window into the Jurassic flora came from the coastal cliffs of North Yorkshire, England. Layers of rock standing like a Victoria sponge cake above the crashing waves recorded the rise and fall of sea level for the entirety of the Jurassic. During intervals when sea level was relatively low, the rocks preserved plant-bearing sediments deposited near the mouths of ancient rivers. The best-preserved plants come from moments when the ancient rivers flooded, and blankets of sand washed across the landscape, burying the forests as they stood. In an approximately 100 km (62 mi) stretch of seacoast from Middlesbrough to Bridlington, nearly 600 fossil-bearing layers have been described. The three most productive, the Whitby Plant Bed, the Gristhorpe Plant Bed, and the Scalby Plant Bed, each named for nearby villages, have yielded more than 250 species.

The canyonlands of Utah paint a different picture of Jurassic-age plants. During the Early Jurassic, the region was covered in a sea of sand dunes, now preserved as the Navajo Sandstone. During rainy periods, a carpet of lichen and dwarf ferns stabilized the dune crests, and beachball-shaped Bennettitales grew in the sandy soil. Low places between the dunes filled with long, narrow lakes teeming with tiny shrimp. Fern and horsetail marshes lined the lakes, and stands of conifers, with cycads in their shade, stood back from the water's edge. The conifers grew without seasonal rings, indicating that the temperature and rainfall remained constant throughout the year. In time, the climate dried again and the sands rolled over and preserved these oases. In the Late Jurassic, the climate became wetter and the Morrison Formation in Utah, Wyoming and Colorado, preserved a wide range of plants and animals that roamed the forested floodplains.

CRETACEOUS PLANT FOSSIL HOT SPOTS

Clays of the Early Cretaceous Lusitanian Basin of western Portugal have yielded extraordinary insights into the early evolution of flowering plants. During the Barremian and Aptian ages of the Early Cretaceous, wildfires on the landscape singed tiny flowers blooming along pond margins. Turned to charcoal, the flowers and seeds wafted on the fire winds to settle into nearby ponds, where the delicate fossils were encased in clay and preserved. Because the sediment was never deeply buried, the clays remain soft, and a gentle rinse releases a bounty of tiny fossils. To an untrained eye, these fossils look like blackened specks, but under an electron microscope they revealed astonishing detail. Although they will never wow visitors in a museum display, dozens of species have been recognized from thousands of minute specimens. Many belong to living angiosperm groups that evolved close to the origin of flowers. The Lusitanian Basin clays also preserve mosses, lycophytes, conifers, Bennettitales, and Gnetales, all bearing witness to a diverse forest ecosystem coping with a drying climate.

Across the Atlantic, similar, but younger, clays in the Potomac Group in Virginia and Maryland record the next phase of the flowering plant radiation. The oldest beds preserve pollen and leaves of a handful of species with relatively simple forms. Over time, new and increasingly complex species appear. Detailed analysis of the sedimentary rocks showed that the Potomac Group plants grew on the unstable banks of ancient rivers. When James Doyle and Leo Hickey published these analyses in 1977, the insights revolutionized the way paleobotanists thought about the early evolution of flowering plants. Previously, most paleobotanists imagined early angiosperms as stately trees with large, *Magnolia*-like flowers. This was the first hint that the radiation of the flowering plants was linked to features shaped by life in disturbed and ephemeral habitats. Lake sediments of the Yixian Formation in Jinzhou, Liaoning Province, China, and amber—fossilized resin—from Myanmar provide other important windows into the Early and mid Cretaceous.

Late Cretaceous fossil plants have been described from every continent of the globe. Paleobotanists have

Sea cliffs on the coast of Yorkshire, England, preserve the most detailed and well-studied record of the forests of the Jurassic.

The badlands of eastern Montana, North and South Dakota, and Wyoming
in the USA preserve an extensive record of life in the Late Cretaceous.
Here, plants, dinosaurs, mammals, crocodiles, turtles, fish, and birds were
preserved together, giving paleontologists a detailed look at the ancient
ecosystem. This region also preserves evidence of the bolide impact that
triggered the end-Cretaceous extinction.

discovered rich and diverse plant fossil assemblages from the Amur River region of Siberia and China. The Arctic regions of Russia and the North Slope of Alaska preserve snapshots of the deciduous forests that ringed the Arctic Ocean during the Late Cretaceous. Other floras are scattered across Europe, especially in the Czech Republic, France, Italy, Portugal, and Spain. These record a gradual decline in cycads and Bennettitales, with conifers retreating to coastal forests while flowering plants and ferns dominated freshwater swamps, floodplains, and riverbanks. Diverse floras from the southern continent of Gondwana show the evolution of a distinct flora. Fossils from Patagonia in temperate latitudes and Colombia in the tropics reveal the antecedents of the distinctive ecosystems that exist there today. In Antarctica, fossils from King George Island and James Ross Island yield clues to the vegetation at southern polar latitudes. A submerged forest that is today 32 km (20 mi) offshore in South Africa and rich amber deposits in Ethiopia provide hints of a wet, forested African continent. A handful of fossil plants have been described from Late Cretaceous dinosaur sites in the Winton Formation of central-western Queensland, Australia. These document the ancestors of the Australian rainforest. The South Island of New Zealand preserves an extensive record of Late Cretaceous temperate forests. And a succession of fossil localities from throughout the Late Cretaceous of India reveal ecological and evolutionary responses to the subcontinent's transit from polar latitudes in the Southern Hemisphere, across the equator, to the subtropics in the north.

In addition to all of these, western North America yields among the most comprehensive and well-studied records of Late Cretaceous vegetation. The western continent of Laramidia extended from the subtropics to the Arctic Circle, crossing a broad range of Cretaceous climates. Ongoing mountain-building along the continent's west coast maintained highlands that provided a continuous supply of sediment throughout the Late Cretaceous. Laramidia offered ideal conditions to record changes in vegetation across both space and time. Dakota Group sandstones, mudstones, and clays from Alberta to Arizona preserved coastal and floodplain ecosystems from the Albian through the Cenomanian. These rocks record a shift into hothouse climate with brief hyperthermals that may have accelerated the northward migration of angiosperms and put pressure on the conifers, cycads, ginkgos, and Bennettitales they replaced. Detailed study of pollen and spores from across the region showed the steady northward march of flowering plants and the handoff of dominance from the conifers of the Early Cretaceous to the angiosperms that grew from subtropics to pole at the end of the Mesozoic. As the hothouse subsided and the Cretaceous drew to a close, rocks from the Lance and Hell Creek Formations documented floodplain forests across the region. Under drier climates, fern savannas developed between the gallery forests that lined the great rivers. And in middle to high latitudes, angiosperm-dominated forests and conifer-rich swamps created a patchwork of habitats across the landscape. And finally, localities from the Big Bend in Texas to the Denver Basin in Colorado through Wyoming, Montana, the Dakotas, and Alberta preserve rocks from the moment that the Mesozoic ended 66 million years ago with the impact of the Yucatán asteroid.

USING THE GROUP AND SPECIES DESCRIPTIONS

NAMING FOSSIL PLANTS

Paleobotanists name fossil plants using the rules established for living plants, with a few special provisions that account for how plants are preserved as fossils. To be formally recognized, a plant must be described in writing and the description published in a scholarly journal or book. Plant descriptions start with a name. Carl von Linné formalized modern botanical naming in his 1753 *Species Plantarum*. Linné gave each species a two-word name—a binomial. This was not an entirely new concept. A century before, Caspar Bauhin pioneered replacing sentence-like names with a shorter version that began with a single name and a one- or two-word description. Linné simplified this approach further by consistently using a single name followed by a one-word "trivial" term that pointed to a unique species within a larger group of similar plants. Over time, this approach was codified into the binomials we use today. The first word in the name signifies the genus and, when combined with the trivial designation, becomes the name of each unique species. Of course, Linné wrote all of his names and descriptions in Latin, the language of science in his day. Using Latin as a language of communication across native tongues was so embedded in the scientific work of the eighteenth century that Linné Latinized his own name to *Carolus Linnæus* or, as he is commonly known today, Carl Linnaeus.

PLANT PARTS, FOSSIL TAXA, AND A "WHOLE-PLANT CONCEPT"

When botanists name a living plant, they write their descriptions from a specimen that includes the stem, leaves, wood if the species makes it, and all of its reproductive parts—the whole plant. This ensures that later botanists can reliably assign isolated leaves or flowers to the described species. Unfortunately, plants tend to fall apart when they enter the fossil record, making it rare to find fossils in which all of the parts remain connected. Early paleobotanists came from a botanical tradition in which naming plants was the central job. Therefore, eighteenth- and nineteenth-century paleobotanists developed the practice of using the same botanical rules to name only one part, say leaves or cones. They called such isolated parts "form species." With more discoveries, paleobotanists realized that different form names had been given to separate parts of the same plant. For example, Adolphe-Théodore Brongniart named the pollen cone of one Jurassic cycad from Yorkshire *Androstrobus wonnacotti*. Hugh Thomas and Tom Harris called one specimen, in which several seeds were found attached to a slender stalk, *Beania mamayi*. Harris later linked both

of these reproductive structures to leaves called *Nilssonia tenuinervis*. However, the stem remained elusive. Harris provided an illustration showing the connection of pollen and seed cones to leaves, but noted that the stem that bore them was, in his words, "imaginary." Although Harris declined to give the "whole plant" a name, the rules of botanical naming would require the first part named to take so-called priority and become the name of the whole plant. In this case, the reunited plant would be *Nilssonia tenuinervis* after its foliage. However, other species might have borne this leaf, which is most definitely against the rules. Harris suspected that some specimens of *Nilssonia tenuinervis* might belong to the cycad-like Bennettitales. He sidestepped the whole conundrum by not naming his whole plant.

Students new to paleobotany find the form-species and whole-plant-species kerfuffle confusing and tiresome. They ask why not jettison the form species names as soon as all the parts have been reunited and the whole plant revealed. Simply put, most fossil plants are described as form species, with only a very few later linked to other form species into something that more closely resembles an entire plant. Retaining the form species names connects the newly reconstructed plant to all of the other instances where its parts were found in pieces. Most importantly, retaining the confusing collection of form names opens the very real possibility that a single leaf form or wood type might be found with different reproductive structures that, botanically speaking, would constitute different species. So, paleobotanists grudgingly tolerate the proliferation of form species names as they search for the ancient plant that produced them.

Most of the names presented in this field guide are form species. Whenever possible, parts are gathered together to make up whole plants, and we describe the confidence with which we make those connections. Sometimes a collection of plant parts may be united because they are always found together, even though they have not yet been found physically connected. This is the weakest association. In other cases, similar features like glands or hairs that appear on various parts can connect form species. In the best case, we find leaves and wood and reproductive structures connected in a single specimen.

The names given to plant parts provide clues to the parts they represent. For example, names for wood generally end in *-xylon*. Leaf names are commonly *-phyllum* or *-phylloides*. Cones may be *-strobus*; other pollen-bearing structures may be *-anthus*. Seeds and fruit sometimes end in *-carpon*, *-carpum*, or *-carpus*. Spore names generally end in *-sporites*, and pollen names end in *-pollis* or *-pollinites*. Although these conventions can help, sometimes things named *-sporites* later turn out to be pollen. In other cases, these helpful clues may not have been provided by the paleobotanist who named the fossil in the first place, and one is left to read the description in detail to understand to what part the name applies.

UNDERSTANDING EVOLUTIONARY RELATIONSHIPS

Without an evolutionary scaffold, little about the history of life makes sense. Plants too are best understood in the context of their evolutionary relationships. Paleobotanists visualize evolutionary relationships with branching diagrams called phylogenies. These are constructed using three simple principles that express our understanding of how evolution works. First, lineages change through time. Mutations accumulate in living individuals and are passed down to their offspring generation after generation. Some of these mutations produce new features that will be tested by the environment during natural selection. Second, any group is related to others by descent from a common ancestor. In many cases, the identity of the common ancestor remains unknown, but it did exist at some point, even if no trace of it remains. Third, at the point when new species arise, lineages diverge and begin independent evolutionary journeys that are distinct from each other and from their common ancestor. For animals, this rule is absolute. For plants, their astonishing ability to produce offspring across species—hybrids—means that divergent lineages do not always remain independent. At the scale of large groups like conifers, cycads, or flowering plants,

the idea of independent branches in the tree of life accurately reflects what happened in nature. However, at the level of species and sometimes even genera, plants create a weblike pattern of evolutionary relationships rather than a neatly dichotomizing tree.

Botanists studying living plants use their genetic material—DNA and RNA—to reconstruct evolutionary relationships. In species known only from fossils, DNA is almost never preserved. Therefore, paleobotanists must rely on the physical features of the plant. However, because there are only so many possible solutions to the challenges of being a plant on land, this task is difficult. Among Jurassic plants, for example, a variety of unrelated lineages independently experimented with elaborate reproductive structures that wrapped their delicious ovules in parental tissue and made a few leaves more colorful to attract pollinators. This phenomenon, called convergent evolution, can complicate attempts to unravel evolutionary relationships from plant features alone. The most notorious example involves the living group that includes the tropical genus *Gnetum*. *Gnetum* has a variety of features reminiscent of flowering plants: large, water-conducting tubes, broad, net-veined leaves, and flowerlike reproductive structures. This led to the suggestion that *Gnetum*, plus the Bennettitales, shared a close common ancestor with flowering plants. Interpretation of the physical features of these groups made a convincing argument. However, about a decade later, DNA sequencing revealed that *Gnetum* was most closely related to pines and other conifers, and flowering plants were only distantly related to all of the remaining seed plants. Thankfully, many common Mesozoic lineages left survivors in the modern flora and can contribute their DNA to unraveling the big picture of Mesozoic plant evolution. And those plants for which we only have fossils can be placed in the larger framework based—as best we can—on the features they possess. For the paleobotanist, humility is essential. Tomorrow, someone may discover a new fossil that changes everything we thought we knew.

INFORMATION PROVIDED WITH EACH ENTRY

The Mesozoic flora contained many more species than could be profiled in this guide. Species were chosen as representatives of the breadth of plant diversity and to highlight as many geographic regions and environments as possible. Some species were also chosen because they have a particularly interesting story to tell about our understanding of plants of the past. Others have historical significance. For each entry in this field guide, we include several pieces of information. **LOCATION** describes where the fossil was found presented in modern geography. **AGE** provides the geological age of the rocks in which the species was initially found. **CHARACTERISTICS** discusses the key features that distinguish this species from others. **FOUND IN** describes the fossil's geological context, including the rock formation and type in which it was found. **HABITAT** reconstructs the aspects of the physical and biological environment that can be gleaned from the rock record. And the **NOTES** section includes important or novel aspects of the plant. For example, the species might have particular historical importance, an interesting evolutionary history, or be a paleobotanical celebrity for other reasons.

GROUP AND SPECIES DESCRIPTIONS

PLANTS THAT REPRODUCE WITH SPORES

In the spore-producing lifestyle, the green leafy plant usually has two copies of each gene tucked within the nuclei of its cells. When this sporophyte is ready to reproduce, special cells divide without first making copies of its DNA to produce spores. Now with just one copy of each gene, the spores are coated in the biopolymer sporopollenin and sent into the world as spores (see pp.36–37). In the soil, spores germinate into tiny, inconspicuous plants—gametophytes, so-called because they produce gametes—that will live independently for some time before producing egg cells and sperm. Sperm swim through water in the environment, drawn by chemical signals produced by the egg cell. When they meet, egg cell and sperm each contribute one copy of the plant's genes to grow a new sporophyte.

Spore-producing plants have only small, simple water-conducting tubes (tracheids), and while most have roots, they rely on all-purpose stems called rhizomes to hold themselves in place, to move through the substrate, and to grow upward or climb. Some have large, complex leaves (like ferns), some have simple leaves fed by a single, vascular strand (like lycophytes), and others have inconspicuous leaves (like horsetails).

Because the sperm of spore-producing plants must swim through the environment in search of the egg cell, these plants prefer wetlands, moist climates, and the humid understory of forests where water is consistently available.

ISOETALES

Small members of this group grow worldwide today. During the Paleozoic, tree-sized forms also evolved but perished during the Late Carboniferous plant extinction. Among the group's Mesozoic representatives, compact stems typically ended in a protective swollen base (corm) that stored carbohydrates. The corms helped the Isoetales survive the end-Permian extinction and subsequently made them important members of the earliest Triassic floras.

MORPHOLOGICAL FEATURES Members of the Isoetales have a squat stem with slender leaves arranged in a tight spiral and attached to the stem by thickened bases. When leaves fall off

OPPOSITE: *Pleuromeia sternbergii* grew to 2 m (6.5 ft) in height.

the stem, they leave characteristic polygonal scars. All lycophytes—Isoetales, Lycopodiales, and Selaginellales—have leaves with only a single, unbranched strand of tracheids per leaf. This feature limits leaves to small, spiky forms or long, narrow blades. In Isoetales, the base of the stem swells into a corm, studded with unbranched roots. Kidney-shaped structures called sporangia, which open like small change purses, hold spores. Sporangia develop on specialized leaves and commonly cluster in a cone. Plants in the Isoetales produce two sizes of spores: microspores (ranging in size from 30–80 μm) that germinate into sperm-producing gametophytes, and megaspores (over 100 μm) that germinate into egg-producing gametophytes.

Annalepis brevicystis

LOCATION Western Guizhou and eastern Yunnan Provinces, China.
AGE Earliest Triassic, Induan (251 million years ago).

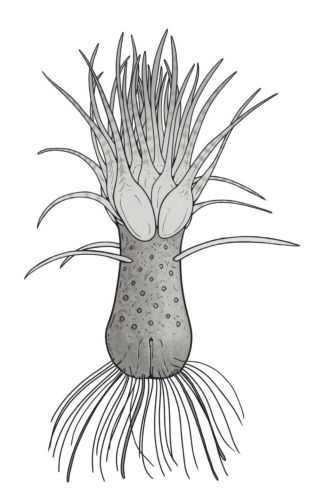

Annalepis brevicystis.

CHARACTERISTICS Small plant with a corm covered in a spiral of sturdy leaf bases. Sporangia developed on leaves smaller than those specialized for photosynthesis. Produced *Aratrisporites*–type microspores.
FOUND IN Marine siltstone and sandstone of the Kayitou Formation.
HABITAT Coastal wetlands where they were easily uprooted and washed into the ocean.
NOTES This species survived the end-Permian extinction and anchored the first communities to recolonize the devastated land. Its slow growth and ability to survive harsh conditions may have given it an advantage.

Isoetites brandnerii

LOCATION Kühwiesenkopf, Italy.
AGE Middle Triassic, Anisian (247–242 million years ago).

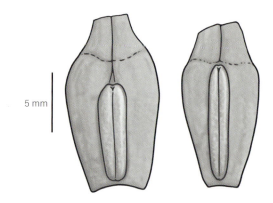

5 mm

Annalepis sporophylls, specialized leaves that held sporangia, were often preserved apart from the plant.

Isoetites brandnerii.

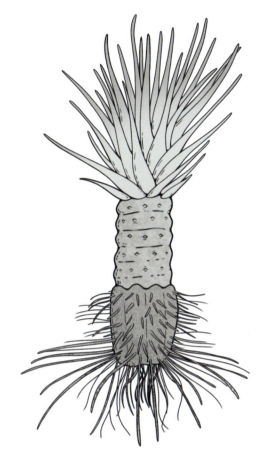

Nathorstiana arborea.

CHARACTERISTICS Short (1–2 cm / 0.4–0.8 in) stem that preserved immature sporangia containing either microspores or megaspores. Leaves up to 8 cm (3.1 in) long, thick, with sharply pointed tips.

FOUND IN Silty and sandy limestone of the Dont Formation.

HABITAT Coastal wetlands.

NOTES Very abundant at this locality, suggesting that *Isoetites brandnerii* exploited a habitat unfavorable for other species.

Nathorstiana arborea

LOCATION Quedlinburg, Germany.

AGE Early Cretaceous (145–100 million years ago).

CHARACTERISTICS Same vascular structure and root arrangement as in living *Isoetes* but much larger, perhaps about 20 cm (7.9 in) tall. Long, thin (a few millimeters wide) leaves attached to the stem with thick bases. Specimens preserved from many individuals at different growth stages show that the base of the stem grew down into the substrate and produced new roots as the aboveground shoot grew upward.

FOUND IN Schlossberg sandstone representing a near-shore marine environment.

HABITAT Coastal wetlands.

NOTES Similarities among the Triassic-age *Pleuromeia*, the Cretaceous-age *Nathorstiana*, and living *Isoetes* led some to suggest that these plants illustrate a trend toward smaller and more specialized plants in this lineage. However, this simple, linear view of evolution only emerges when researchers overlook the variety of plants in formerly diverse groups like this one. In fact, small members of the Isoetales were also present in the Mesozoic, where they were represented by many species.

Pleuromeia longicaulis

LOCATION North of Sydney, New South Wales, Australia.
AGE Early Triassic (251–247 million years ago).
CHARACTERISTICS Single, unbranched stem 30–50 cm (11.8–19.7 in) tall. Narrow leaves about 10 cm (3.9 in) long and with a single vein developed in a tight spiral in the upper 5 cm (2 in) of the stem. Leaves on

the lower part of the stem fell off, leaving characteristic diamond-shaped scars. The base consisted of an unlobed corm with abundant rootlets. The single cone produced at the top of the plant was described under the name *Cyclostrobus sydneyensis*. The cone was cylindrical, about 2 cm (0.8 in) long and 1 cm (0.4 in) wide. The non-photosynthetic leaves that made up the cone were wedge-shaped, about 7 mm (0.3 in) long, with a prominent keel on their outer side. These modified leaves and the sporangia they housed may have dispersed by falling off the cone and floating away.
FOUND IN Sandstones of the Newport Formation.
HABITAT *Pleuromeia longicaulis* grew in extensive single-species stands on the shores of a large coastal lagoon. It may have occupied the brackish zone shoreward of a more diverse vegetation that included the tree-sized horsetail *Neocalamites*, the ferns *Cladophlebis* and *Gleichenites*, and the common Gondwanan seed fern *Dicroidium*.
NOTES *Pleuromeia* evolved in what is today northern China during the earliest Triassic and migrated along the coastlines of Pangaea to reach this part of Gondwana at the end of the Triassic. Ecologically, *Pleuromeia*'s tolerance for harsh conditions allowed it to be among the first large plants to re-green the Triassic landscape after the end-Permian extinction event. This same tolerance for environmental stress allowed it to persist in brackish conditions as other groups pushed it out of freshwater swamps.

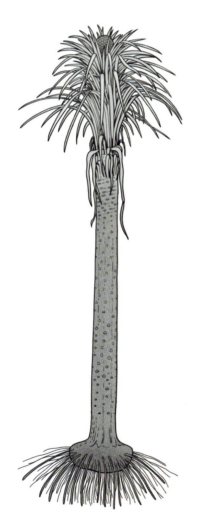

Pleuromeia longicaulis, the plant that produced *Cyclostrobus sydneyensis* cones.

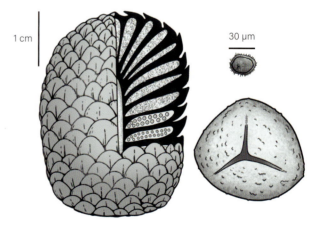

1 cm

30 μm

Cyclostrobus sydneyensis produced large spores in which egg cells developed and small spores that produced sperm.

The coast of Australia in the Triassic. *Pleuromeia* lined the coast, and large *Cyclostrobus* cones were dispersed by the sea.

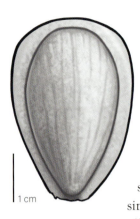

1 cm

The sporangium of *Pleuromeia obovata* developed on a modified leaf.

Pleuromeia obovata

LOCATION Shaanxi Province, northern China.
AGE Middle Triassic, Ladinian. Volcanic ash beds surrounding the fossil date it to 241 million years ago.
CHARACTERISTICS The species was described from a single, spherical cone about 5 cm (2 in) long and 4 cm (1.6 in) wide. It was composed of non-photosynthetic leaves shaped like rounded triangles, with the widest part facing the top. Cone leaves were 1–2 cm (0.4–0.8 in) long and arranged in a tight, overlapping spiral, and each bore a sporangium that was nearly as large as the leaf itself.
FOUND IN Tongchuan Formation of the Ordos Basin.
HABITAT Swampy floodplain.
NOTES This is the youngest *Pleuromeia* described to date. The single cone was preserved in a volcanic ash bed along with isolated sporangium-bearing leaves from the same species. The plant was destroyed during the eruption that preserved its fossil.

Pleuromeia rossica

LOCATION Tikhvinskoe locality near Rybinsk, northern Volga region of Russia.
AGE Early Triassic, Olenekian (251–247 million years ago).
CHARACTERISTICS Unbranched stem ranging from 50–60 cm (19.7–23.6 in) tall and 2–3 cm (0.8–1.2 in) wide. Photosynthetic leaves rare and when present widely spaced around in the upper third of the plant. The corm of the plant was globular, with two to four lobe-like extensions that helped anchor the plant in the unstable substrate. The corm was covered in root scars about 1 mm (0.04 in) in diameter. The single cone produced at the top of the stem contained both small (described as dispersed spores under the name *Densoisporites neuburgae*) and large spore types. Fully developed cones were 7–8 cm (2.8–3.1 in) long and 5 cm (2 in) wide. The non-photosynthetic leaves that

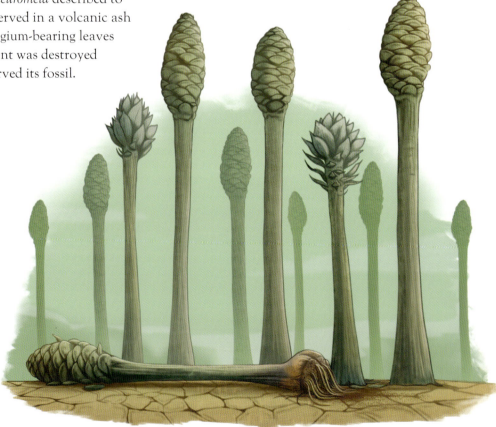

Pleuromeia rossica grew in dense, single-species stands.

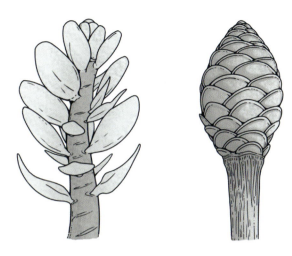

Pleuromeia rossica produced photosynthetic leaves (left) and sporangium-bearing leaves packed into a cone (right).

made up the cone were oval and arranged in a dense, overlapping spiral.

FOUND IN Sandy mudstone of the Rybinsk Formation.

HABITAT *Pleuromeia rossica* tolerated brackish conditions and grew in dense stands in tidal mudflats. Many of the fallen stems lay in roughly the same direction, suggesting that they were knocked down by storm waves.

NOTES Whole *Pleuromeia rossica* plants at many life stages are preserved at Tikhvinskoe with aboveground stems still attached to corms.

Pleuromeia sternbergii

LOCATION Bernburg, Germany.

AGE Early Triassic, Olenekian (251–247 million years ago).

CHARACTERISTICS A single, unbranched stem could grow up to 2 m (6.5 ft) tall. Slender leaves about 10 cm (3.9 in) long that tapered toward the tip and emerged in a tight spiral from the stem. A single, unbranched vein extended the length of each leaf. Corm, with or without lobes, from which roots grew. A single cone developed at the tip of the plant. Megaspores of the *Densosporites* type were produced in sporangia in the lower part of the cone. Microspores of the *Trileites* type developed at the top. Because the plant did not

branch, it produced only one cone at the end of its life.

FOUND IN Bunter sandstone. The first specimen was found when a block of rock from Magdeburg Cathedral fell and split open, revealing the fossil. More specimens were later found in quarries of this building stone.

HABITAT *Pleuromeia sternbergii* grew in single-species stands on poorly drained soils. It likely grew slowly for most of its life and survived in places that other plants could not exploit.

NOTES This species is among the most common of all the *Pleuromeia* and it has been reported from a variety of locations, including Germany, Spain, France, Russia, and China.

Pleuromeia sternbergii with a spore-producing cone on top.

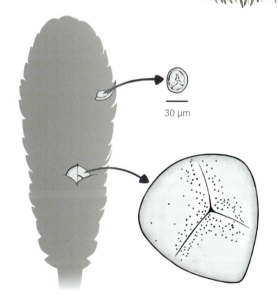

Pleuromeia sternbergii cones produced both large spores that produced egg cells and small spores that produced sperm.

30 µm

LYCOPODIALES

This group evolved early in the history of land plants and diverged from the other spore-bearing plants about 400 million years ago. Although Lycopodiales was probably abundant in the Mesozoic, its members were small plants that occupied environments where they were unlikely to enter the fossil record, such as forest understories.

MORPHOLOGICAL FEATURES Today, members of the Lycopodiales are small, herbaceous plants. Some Mesozoic representatives may have grown a bit taller, but none approached tree-size. Stems branch and are covered in small leaves arranged in a spiral to completely cover the stem. In the stem, tracheids cluster in a single strand at the center. Kidney-shaped sporangia may be scattered along the stem or grouped at a growing tip in a distinct cone-like structure. Sporangia open like a change purse to release a single type of spore, typically with the peace sign-shaped trilete mark on its surface. Small, dichotomously branching roots emerge from anywhere along the stem.

Lycopodites falcatus

LOCATION Runswick Bay, Cayton Bay, and Gristhorpe Bay, Yorkshire, England.

AGE Middle Jurassic, Bajocian to Callovian (170–164 million years ago).

CHARACTERISTICS Small plants that may have hung pendulous from larger trees or grown erected on the ground. Slender stems branched dichotomously in a single plane. Scale-shaped microphyll leaves covered the entire plant. Larger leaves emerged on either side of the branch, with smaller ones above and below. Tracheids formed a single, central bundle. No sporangia have been found attached, but specialized cone-like structures have been found nearby.

FOUND IN Preserved as compressions in siltstone and sandy shale.

HABITAT Floodplain forest, perhaps coastal.

NOTES *Lycopodites* could be considered a "wastebasket" taxon into which many small lycophytes can be tossed. Specimens from Cretaceous sediments of Siberia and India and the Jurassic of Australia, France, Kazakhstan, Mongolia, New Zealand, and Tajikistan have all been placed in the genus. This means the fossils resemble *Lycopodites* but preservation is not good enough to assign them to a previously described species or to erect a new one. In addition, paleobotanists have described many species from around the world and throughout the Mesozoic. *Lycopodites sahnii* comes from the Triassic of Madhya Pradesh, India, *L. scanicus* from the Early Jurassic of Sweden, *L. victoriae* from the Jurassic of Australia, *L. arberi* from the Jurassic of New Zealand, *L. sewardi* from the Mesozoic of Norway, *L. multifurcatus* from the Early Cretaceous of northeastern China, *L. ghoshii* from the Early Cretaceous of India, and *L. tuscaloosensis* from the Late Cretaceous of Alabama, USA. Plants with the same basic features persist today, with about 400 described species.

Lycopodites falcatus.

SELAGINELLALES

This group of lycophytes first appeared in the Carboniferous and grew as a creeping herb alongside the tree-sized Isoetaleans. Their spores can survive for long periods dormant in the environment, which may have helped them survive both the Late Carboniferous plant extinction and the end-Permian extinction.

MORPHOLOGICAL FEATURES Plants grow from a rhizome that creeps through the leaf litter, sending up vertical shoots at intervals. Leaves of different sizes cover the upright shoot, while the creeping rhizome remains bare. Leaves on the top and bottom of the shoot are small and scalelike. Leaves emerging from the sides are larger and their shape varies among species. This arrangement of leaves produces a flattened, light-catching surface. Like members of the Isoetales, Selaginellales produce two different sizes of spores. Sperm-producing gametophytes germinate from the smaller spores and live independently for a time in leaf litter before releasing sperm into the environment, while egg-producing gametophytes complete their development and invite fertilization within the walls of the larger spores. This feature allows megaspores to survive extended periods of drought.

Selaginella cretacea

LOCATION Hukawng Valley, Myanmar.

AGE Cretaceous, Albian to Cenomanian (113–94 million years ago).

CHARACTERISTICS The edges of all leaves were slightly serrated. On the top of the stem, leaves were asymmetrical and oval; on the bottom of the stem, they were symmetrical, overlapping, and oval to elongate in shape. A single strand of tracheids entered each leaf. The sporangia were elliptical and developed on non-photosynthetic leaves clustered into a cone-like structure at the tips of branches.

FOUND IN Amber, the fossilized resin of many plants but mainly conifers and tree ferns.

HABITAT The abundance of amber at this site suggests a forest with a dense overstory of resin-producing conifers and tree ferns. *Selaginella cretacea* probably grew on tree trunks, which would have enhanced the likelihood of being entombed in resin.

NOTES This Cretaceous-age representative is similar enough to living species to be placed in the same genus as all modern species of the group. *Selaginella* was diverse in the Cretaceous, with at least 20 other species described from the Kachin amber of Myanmar alone.

Selaginella cretacea. Sporangia are shown in association with small leaves at branch tips.

Selaginella anasazia

LOCATION Arizona, USA.

AGE Late Triassic, Norian (227–208 million years ago).

CHARACTERISTICS Branching stem in which one branch tended to continue in the same direction and the other extended at an angle to the side. The leaves on the sides of the stem were larger (1 mm / 0.04 in wide and 2 mm / 0.08 in long) and spreading, compared to the smaller, elongate leaves (up to 1 mm / 0.04 in long) that appeared on the top and bottom of the stem. The stem had two bundles of tracheids that divided into four just below each dichotomy. The cone was at least 3 mm (0.1 in) long, with sporangia developed in a spiral on non-photosynthetic leaves. Larger spores appeared on the bottommost part of the cone; smaller spores were produced at the tip.

FOUND IN Mudstone and sandstone of the Chinle Formation.

HABITAT Understory of a forest growing on a river floodplain.

NOTES *Selaginella anasazia* likely flourished during periods of wet climate in the otherwise dry Late Triassic of the region.

Selaginellites leonardii

LOCATION Kühwiesenkopf, Italy.

AGE Middle Triassic, Anisian (247–242 million years ago).

CHARACTERISTICS Branching stem up to 5 cm (2 in) high with branches arising at irregular intervals. Upper and lower leaves small and pressed tightly to the stems; lateral leaves elongated or sickle-shaped. Cone of smaller, non-photosynthetic leaves held sporangia. Large spores produced at the base of the cone were circular. Small sporangia had not yet developed on the preserved specimen, suggesting that they matured later to prevent sperm and eggs from the same plant combining.

FOUND IN Silty and sandy limestone of the Dont Formation.

HABITAT Understory of conifer, fern, and cycad forests.

NOTES This species was not placed in the modern genus *Selaginella*. Instead, authors named the plant using the standard "fossil" form, *Selaginellites*. Some researchers choose to always use the *-ites* suffix to indicate that they are describing an extinct species. Others prefer to assign fossils to living groups whenever they can observe key diagnostic features that match those of living plants. Most do both, depending on the confidence they have in the connection between ancient and modern plants.

Selaginella anasazia.

EQUISETALES

Horsetails arose in the Paleozoic, evolved significant diversity, and survived both the Late Carboniferous plant extinction and the end-Permian mass extinction to become important elements in the Triassic flora. Through the Mesozoic, they lost much of their ecological diversity to persist as small and mid-sized plants that today specialize in sandy wetlands and roadside ditches.

MORPHOLOGICAL FEATURES Horsetails have a strong node–internode architecture. Although all land plants are built from such node–internode modules, the horsetails simplified the design. Nodes are distinct, evenly spaced separations in the stem from which branches and leaves emerge. Internodes are long and straight, commonly with no features other than the fine, vertical ribs that characterize this group. The interior of most horsetail stems is hollow or filled with spongy cells that decay in older stems. Horsetails are commonly preserved when sand fills this hollow interior space. One or many branches grow from any node. The green stem of the plant did most of the work of photosynthesis in Mesozoic Equisetales, leaves being reduced or absent.

Equisetites arenaceus

LOCATION Baden-Württemberg, Germany.
AGE Middle Triassic, Ladinian (242–237 million years ago).
CHARACTERISTICS Straight stems with strong node–internode architectures up to 20 cm (7.9 in) in diameter and 5 m (16.4 ft) tall. Vertical stems emerged from a stout rhizome that produced both roots and upright shoots at its nodes. Whorls of slender branches emerged from most nodes on the upright part of the stem. Some branches, particularly near the base of the plant, produced a whorl of smaller branches at their nodes. Each vertical stem node sported up to 100 leaves that fused into a tight skirt encircling the node. Three cone-like structures containing sporangia developed on the tips of some short branches. Spores were 50–60 μm in diameter, rounded, with thin walls and a small trilete mark.
FOUND IN Silty sandstone.
HABITAT Sandy riverbanks.
NOTES The species was also found in younger, Carnian-age rocks of the same region.

30 μm

Equisetites arenaceus produced smooth spores about 60 μm in diameter.

Equisetites arenaceus with a whorl of spore-producing cones growing from a node.

Equisetites muensteri

Equisetites muensteri with spore-producing cones.

LOCATION Widespread in Europe and Scoresby Sound, eastern Greenland.

AGE Triassic to Early Jurassic (251–174 million years ago).

CHARACTERISTICS Slender, unbranching stems up to 1 cm (0.4 in) in diameter that grew from a rhizome. Variable distance between nodes, with strong, vertical ribs connecting nodes. Approximately 12 leaves grew from each node, fused, and encircled the stem. Stomata for gas exchange developed on leaves and scattered along the stem, demonstrating that both stem and leaves were important for photosynthesis. Clusters of spore-producing cones developed along each vertical stem. Spores were spherical or oblong, with thin walls.

FOUND IN Silty sandstones.

HABITAT Sandy streamsides and sand bars in rivers.

NOTES Except for two characteristics, *Equisetites muensteri* could fit comfortably in the living genus *Equisetum*. First, this species produced cones at many nodes. Modern *Equisetum* produces cones only at branch tips. Second, the spores of all living *Equisetum* have slender projections from the surface that may be four or five times the diameter of the spore itself. Inside the sporangium, these structures coil tightly. When water hits the spore, the coil unfurls rapidly and slings the spore away from the parent plant. All members of *Equisetum*—living and fossil—have this feature; *Equisetites* did not.

Neocalamites merianii

LOCATION Baden-Württemberg, Germany.

AGE Middle to Late Triassic, Ladinian to Carnian (242–227 million years ago).

CHARACTERISTICS This horsetail had strong node–internode architecture with vertical stems up to 2 m (6.5 ft) tall and up to 10 cm (3.9 in) in diameter. Slender branches emerged from each node to form spiky rings around the stem. Stems and branches had strong, parallel ribs of water-conducting cells running vertically from node to node. A whorl of long, narrow, single-veined leaves emerged from some nodes. These leaves, which could be up to 15 cm (5.9 in) long, supplemented photosynthesis that mostly occurred on green stems and branches. Little is known about the underground parts of *Neocalamites*, but it probably followed the horsetail model of a rhizome with rootlets emerging at the nodes. *Neocalamites* reproduced by spores that developed in cones that grew from nodes.

FOUND IN Silty sandstone.

HABITAT Sandy margins of streams and rivers.

NOTES The name *Neocalamites* means "new *Calamites*" and applies to plants that resemble the tree-sized horsetail of the Paleozoic swamps, *Calamites*. *Calamites* was common in the equatorial coal swamps and perished during the Late Carboniferous plant extinction. *Neocalamites* appeared in the Permian and resembled its predecessor in many ways, except that *Neocalamites* did not produce wood.

Neocalamites merianii with whorls of branches emerging from nodes.

MARATTIALES

This group, among the most ancient of the true ferns, was abundant and diverse in the Paleozoic and included herb- and tree-sized species. Marattiales lost significant diversity in the Late Carboniferous plant extinction and never recovered its earlier grandeur. However, it remained a persistent and distinctive member of wet tropical understory communities throughout the Mesozoic and today includes 110 living species worldwide.

MORPHOLOGICAL FEATURES Large, two- or three-times pinnate leaves. Stem may be slender and creeping or short and stout, with a growing tip that produces one cluster of leaves at a time. The leaf stem, also known as the petiole, can be meters long, allowing leaves to reach toward the canopy. Base of the petiole swollen and fleshy. Sporangia develop from several cells on the surface of the leaf and cluster together in a sturdy, protective structure.

Marattia intermedia

LOCATION Ghoznavi region, Iran.
AGE Triassic (251–201 million years ago).
CHARACTERISTICS Whole fronds of this group were rarely preserved because of their extraordinary size.

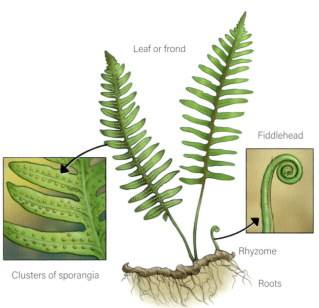

Parts of a fern plant.

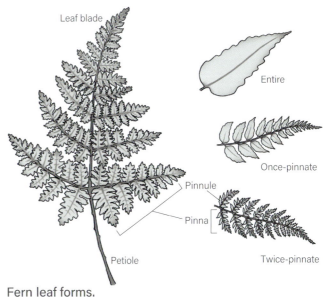

Fern leaf forms.

Marattia intermedia. Trunks could grow into small trees.

Most fossils were isolated parts of the leaf with or without sporangia. Individual leaf segments up to 50 cm (19.7 in) long and those with sporangia were identical to those without. A dominant vein ran down the center of each pinnule, with smaller, lateral veins extending from it toward the edge and forking in some larger examples. Sporangia-containing structures elongate and boat-shaped, much like those

103

of modern species. Spores produced within opened from a single slit.
FOUND IN Organic-rich shale.
HABITAT Wet understory forested floodplain in deep shade.

NOTES *Marattia intermedia* was assigned to the modern genus because it shared all the characteristics of its living members, including a preference for wet, shady, understory habitats.

OSMUNDALES

These understory ferns originated in the Late Permian and survived the end-Permian mass extinction. Never very diverse or common in the fossil record, the Osmundales are sister to the main diversity of ferns and are found worldwide from the Mesozoic to today. Osmundales ferns live in a wide range of climates and habitats because their rhizomes and spores can survive cold and dry conditions.
MORPHOLOGICAL FEATURES Large leaves that may be once- or twice-pinnate. Leaves covered in hairs or scales that deter herbivores and slow water loss. Leaves arise in a spiral from the growing tip of a short stem. Tracheids in the stem arranged in a ring of bundles that divide and rejoin along the length of the stem. Vascular strands that deliver water to the leaves diverge from one of these central bundles at intervals. Sporangia develop from a single cell on a photosynthetic leaf or on highly modified leaves that stand spike-like above the photosynthetic leaves. Sporangia have thin walls with a zone of large, thick cells that pull the sporangia apart at maturity, dispersing spores.

Ashicaulis hebeiensis

LOCATION Hebei Province, China.
AGE Middle Jurassic (174–163 million years ago).
CHARACTERISTICS Tracheids in the stem arranged in 9 to 14 small bundles embedded in a mass of rounded, thick-walled cells that gave the stem support. Stem fossils found with once-pinnate leaves named *Cladophlebis delicatula* and *Todites* for fertile forms. Since neither of these leaf types were attached to the stem, paleobotanists cannot be certain whether one or both were produced by *Ashicaulis hebeiensis*–type stems.
FOUND IN Xiahuayan coal mine shale beds.
HABITAT Swamp forest.
NOTES Originally described as *Osmundacaulis hebeiensis* in 1983, the genus *Osmundacaulis* was split into two genera—*Millerocaulis* and *Ashicaulis*—by Mesozoic fern expert Don Tidwell. Tidwell retained *Osmundacaulis* for tree-sized forms and dedicated the

Ashicaulis hebeiensis is known from its underground stem. Leaves shown are an interpretation.

Cross section of the *Ashicaulis hebeiensis* stem.

two new genera to species with a creeping rhizome or a short, upright stem. *Ashicaulis hebeiensis* is one of very few instances where the stem and leaves were found together.

Cladophlebis denticulata

LOCATION Worldwide distribution including North America, Europe, Asia, South America, and the Middle East.

Cladophlebis denticulata.

AGE At least Triassic through Early Cretaceous (251 to 100 million years ago).

CHARACTERISTICS Leaves twice-pinnate with the smallest pinnae arranged opposite one another or alternating on either side of their support structure. Pinnae attached to the leaf's support structure across their entire width and were pointed, with slightly scalloped edges. A single vein entered each pinnule and extended to its tip. Smaller veins branched from the major vein and bifurcated as they got closer to the edge of the pinnule.

FOUND IN Shales and siltstones.

HABITAT Floodplains and possibly disturbed areas.

NOTES *Cladophlebis denticulata* was described by Adolphe-Théodore Brongniart in 1828 and remains one of the most common fern fossils from the Triassic and Jurassic. Foliage features vary, even in the same leaf, which, when fragments of leaves are found apart, has led to a proliferation of species names. When large fronds are found, paleobotanists generally report that their single leaf contains characteristics that define separate described species based on smaller fragments. Where it is found, *Cladophlebis denticulata* tends to be very abundant, suggesting that it might have grown in dense stands, possibly colonizing disturbed areas. The name *Cladophlebis denticulata* is reserved for specimens in which spores are not present. It may be that the photosynthetic frond of *Cladophlebis denticulata* differed considerably from the specialized leaf on which spores developed. Having two very different-looking fronds is common in *Osmunda*–type ferns.

Fossils preserve the distinctive signs of insects eating *Cladophlebis* leaves from the inside, so-called leaf mining.

Osmunda cinnamomea

LOCATION Drumheller, Alberta, Canada.

AGE Late Cretaceous, Maastrichtian (72–66 million years ago).

CHARACTERISTICS Short, upright stem armored with the remains of sturdy leaf bases and fibrous roots that grew from the stem. Twice-pinnate leaves, with pinnules connected to one another at their bases. A central vein entered each lobe and smaller veins branched from it. Sporangia developed on specialized hairy leaves dedicated to reproduction.

FOUND IN Ironstone concretions associated with coal seams.

HABITAT Swamp forest.

NOTES This 72-million-year-old fossil is indistinguishable from its living descendants. Fossil *Osmunda cinnamomea* showed that fern species can be extraordinarily long-lived and that some lineages tend not to change once they have found an environment in which they are successful. The geological context of the Cretaceous fossils suggests that ancient *Osmunda cinnamomea* loved the same marshy forests as the living cinnamon ferns seen across North America today.

Todites princeps

LOCATION Mecsek Mountains, Hungary.

AGE Early Jurassic, Hettangian (201–199 million years ago).

CHARACTERISTICS Twice-pinnate fronds with pinnules about 6 cm (2.4 in) long, slender, generally less than 1 mm (0.04 in) wide. Pinnules were arranged opposite one another on the leaf's support structure or alternated. Pinnules varied in shape, with some on the same specimen having rounded tips and others slightly pointed. Margins were deeply scalloped or almost straight. A single vein entered each photosynthetic lobe and divided repeatedly toward the margin. Sporangia developed in dense clusters on the lower surface of some leaves.

FOUND IN Siltstones.

Cretaceous-age *Osmunda cinnamomea* was identical to its modern descendant.

Todites princeps.

HABITAT Wet, possibly coastal, floodplains.
NOTES *Todites princeps* was found among most of the collections from this location, showing that it was widespread and occupied a diversity of habitats, but only a handful of individual specimens were found at each site, indicating that it was uncommon. This contrasted with *Cladophlebis denticulata*, which was also found at this site. When present, *Cladophlebis denticulata* occurred alone or with leaves of the cycad *Nilssonia*, confirming that it likely grew in dense thickets, while *Todites princeps* was a ubiquitous part of more diverse plant communities.

HYMENOPHYLLALES

Today, filmy ferns live on the ground, attached to trees, or among rocks in shady, wet areas. They are small and inconspicuous. Their minute stems, diaphanous leaves, and forest habitat make this group unlikely to fossilize. Studies of fern relationships show that the filmy ferns, along with Osmundales, evolved early in the history of modern ferns, but only a few fossils corroborate this claim.
MORPHOLOGICAL FEATURES Rhizome is thin and hairy with a single bundle of tracheids in its center. Leaves small, thin—generally only one cell layer thick—sometimes scaly and come in a variety of shapes. Clusters of sporangia develop from a single cell at the margin of photosynthetic leaves.

Hopetedia praetermissa

LOCATION North Carolina, USA.
AGE Late Triassic, Carnian (237–227 million years ago).
CHARACTERISTICS Thin (2.5 mm / 0.09 in diameter), creeping stem that produced three-times pinnate leaves with alternating ultimate lobes. Leaf tissue webbed in between the branching vascular bundles of the leaf to form a fingerlike surface to catch light. Leaves lacked hairs or scales. Leaf margins scalloped, with sporangia developing at the edge of the lobe in the scallop. A funnel-shaped covering

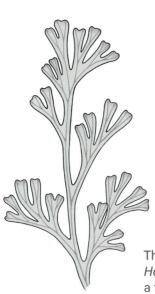

This fragment of the frond of *Hopetedia praetermissa* is only a few centimeters long.

surrounded and protected the sporangia, which grew on a squat stalk.
FOUND IN Clay.
HABITAT Tropical pondside forest.
NOTES Triassic-age *Hopetedia praetermissa* contained a mix of features that, by Cretaceous time, distinguished separate genera. *Trichomanes* had only the funnel-shaped tissue surrounding each cluster of sporangia. *Hymenophyllum* sporangia clusters developed on short stalks but without the funnel-shaped container. This suggests that *Hopetedia* may have been the ancestor of the later genera, each of which lost one of the former features.

Trichomanes angustum

LOCATION Kachin State, Myanmar.
AGE Mid-Cretaceous, late Albian to early Cenomanian (110–95 million years ago).
CHARACTERISTICS Leaves three-times pinnate with the smallest pinnules divided into forked segments and webbed with photosynthetic tissue. Clusters of sporangia surrounded by a funnel-shaped structure

developed on the tip of pinnules.

FOUND IN Amber, the fossilized resin of many plants but mainly conifers and tree ferns.

HABITAT Tropical rainforest.

NOTES Fossils preserved in amber can be extraordinarily

detailed because chemicals in the resin inhibited microbial decay. This was especially important for delicate plants like the filmy ferns. Amber also preserved organisms in three dimensions, rather than squashed flat as in many plant fossils. Consequently, paleobotanists can describe plants in amber with almost the same detail that they might observe in living plants.

This fragment of the frond of *Trichomanes angustum* is only a few centimeters long.

GLEICHENIALES

This fern group originated in the Carboniferous and remained rare through the Permian. In the Mesozoic, Gleicheniales became a common, if not diverse, member of the fern flora. Today, the group is confined to tropical and subtropical climates, but during the warmer Mesozoic, members of the Gleicheniales ranged from pole to pole.

MORPHOLOGICAL FEATURES Long, creeping rhizomes allow plants to explore their environment. When the rhizome discovers a light patch, it pauses growth and forms dense thickets of leaves. The rhizome is hairy or scaly with a single strand of tracheids that branch to feed each leaf. Tracheids in the petiole look circular or C-shaped in cross section. Leaves are at least once-pinnate with sturdy, long-lived, photosynthetic surfaces. Clusters of sporangia develop on the underside of the leaf, generally near the first vein that branches into the smallest lobe. Sporangia grow exposed—without a covering supplied by the parent plant—in clusters on short stalks.

Calcaropteris boeselii

LOCATION Northeastern Washington State, USA.

AGE Mid-Cretaceous, Albian to Cenomanian (113–94 million years ago).

CHARACTERISTICS Described from a complete leaf about 30 cm (11.8 in) long and 15 cm (5.9 in) wide at its widest part. Leaf once-pinnate with pinnules of variable lengths, becoming shorter toward the leaf tip. Long, stout petiole had a distinctly swollen base.

Calcaropteris boeselii.

In some specimens, the petiole was as long as the rest of the leaf, suggesting the ability to stretch into the canopy. Main axis of the leaf tapered toward the tip. Pinnules were up to 10 cm (3.9) long and 1 cm (0.4 in) wide and had scalloped edges with a single, wide vein extending into each. They were arranged in an alternate to subalternate pattern along the main axis of the leaf and diverged at an angle of about 60°.

FOUND IN Winthrop Formation, shallow marine shales, and floodplain deposits with fossil soils.

HABITAT Coastal streamside forest.

NOTES *Calcaropteris boeselii* is the most abundant fern species in the Winthrop Formation. It has been found in a variety of environments but rarely in the forest understory. *Calcaropteris boeselii* likely grew in thickets on frequently flooded areas near the river. The thick base of the petiole may have allowed leaves to break off easily during floods to protect the rhizome from being uprooted.

Gleichenia chalonerii

LOCATION Bedfordshire, England.

AGE Early Cretaceous, Albian (113–100 million years ago).

CHARACTERISTICS Large leaves with characteristic C-shaped vascular strand in the petiole. Some leaves grew from scaly "resting buds" typical of leaves of this group. Pinnules were thick, hairy, covered in thick cuticle, and rolled in at the margins. Stomata sank into the surface of the leaf. Five sporangia clustered together on the underside of the leaf and were surrounded by protective hairs.

FOUND IN Preserved as charcoal in silty shale.

HABITAT *Gleichenia chalonerii* showed several features that suggested it grew in full sun in a periodically hot, dry climate. Small, thick, hairy leaves helped protect the chloroplasts within from intense sunlight. Rolled leaf margins and sunken stomata helped conserve water. Preserved as charcoal, the fossil testified that fire shaped this Early Cretaceous ecosystem.

NOTES Charcoal preserved three-dimensional detail of the plant's internal anatomy.

Gleichenites coloradensis

LOCATION Wyoming, USA.

AGE Late Cretaceous, Cenomanian (100–94 million years ago).

CHARACTERISTICS Large leaves with long petioles. Leaves were four-times pinnate with a trifurcation at each branching point. In each group of three branches, the central branch may have been temporarily prevented from developing. This so-called "resting bud" appeared as a nub between two fully unrolled pinnae. Later activation of this bud allowed the leaf to recover quickly from herbivory.

FOUND IN Non-marine shale.

HABITAT *Gleichenites coloradensis* grew in almost pure stands, suggesting that the species may have colonized open areas after fire, flood, or herbivore disturbance cleared slower-growing plants.

NOTES Botanists consider the name *Gleichenites* "invalid." When a new species is formally named, botanists designate a particular example specimen. They then compare all subsequent candidates for that name with the reference specimen. The genus *Gleichenites* was originally proposed for fernlike foliage that resembled living *Gleichenia*. Unfortunately, most of those fossils did not preserve spores, and later workers realized that they were not ferns at all.

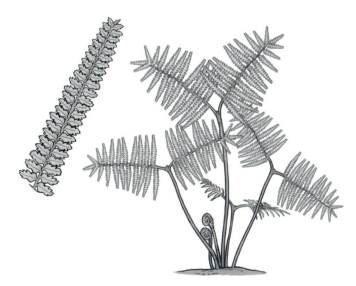

Gleichenites coloradensis.

The name transferred, with the reference specimens, to the new group of plants. Fossils described as *Gleichenites coloradensis* from Wyoming also lacked sporangia but were much more complete and better preserved. They clearly resembled living *Gleichenia*, but because the sporangia remained elusive, Henry Andrews and Cortland Pearsall chose to use the "*Gleichenia*-like" sense of *Gleichenites* for the time being. The name is thus helpful in pointing toward fossils that belong to Gleicheniales, if not technically correct.

Microphyllopteris delicata

LOCATION Northeastern Washington State, USA.
AGE Mid-Cretaceous, Albian to Cenomanian (113–94 million years ago).
CHARACTERISTICS Complete leaves have not been found, but once-pinnate fragments were greater than 10 cm (3.9 in) long and about 8 cm (3.1 in) wide. The central support of the leaf was delicate, less than a millimeter in diameter, with a shallow, V-shaped cross section. Photosynthetic surfaces attached to this central structure were opposite one another, or just subopposite and extended at nearly right angles from the main axis of the leaf. These pinnules were up to 4 cm (1.6 in) long and 3 mm (0.1 in) wide, with scalloped edges and tapered to a point. One vein

Microphyllopteris delicata.

extended into each lobe and divided four or five times to form a series of forks that extended to the leaf margin. Circular clusters of sporangia grouped together in roughly the center of some leaf margin scallops.
FOUND IN Winthrop Formation, shallow marine shales, and floodplain deposits with fossil soils.
HABITAT Coastal streamside forest.
NOTES The combination of traits places this species in the modern family Gleicheniaceae. *Microphyllopteris gieseckiana* was also described from the Winthrop Formation. This much larger frond did not preserve sporangia and had a distinct pattern of veins.

Phlebopteris dunkeri

LOCATION East-central Iran.
AGE Middle Jurassic to Early Cretaceous (174–100 million years ago).
CHARACTERISTICS Only a fragment of the fertile pinnule was recovered from the drill core so nothing is known about the stem or the architecture of the leaf. However, the venation and position of the sporangi were different enough from other described members of the genus to warrant a new name. The fragment was 4.5 cm (1.7 in) long, 1 cm (0.4 in) wide, and very thin. The main vein was depressed into the top surface and raised beneath in a way typical of thick fern leaves. Smaller veins extended from the main vein at right angles and forked once about halfway to the edge of the lobe. After the fork, secondary veins attached to veins from an adjacent branch to form an arc. Still smaller veins emerged from the top of the arc, branched twice, and extended toward the pinnule margin. No sporangia were preserved, but impressions of them occurred about halfway from the main vein and the pinnule margin, suggesting that the spores had already been dispersed at the time the frond was preserved.
FOUND IN Shale within the coal horizons of the Hojedk Formation.
HABITAT Floodplain swamp forest.
NOTES *Phlebopteris dunkeri* has also been reported from the Middle Jurassic to Early Cretaceous of Germany, England, Scotland, France, and Poland. The Iranian fossils were preserved as charcoal, indicating that the species grew in a wildfire-prone climate.

SCHIZAEALES

Ferns in this group grow on the ground from short stems, live as epiphytes on other plants, or climb. Today, they are mostly restricted to frost-free climates. No members of the Schizaeales have been discovered from the Triassic, but by the Late Jurassic, they were globally distributed if uncommon.

MORPHOLOGICAL FEATURES Ferns of the Schizaeales are varied in form, united by having a cluster of thickened cells at the top of their sporangia that, when they dry out, split apart to release the spores inside.

Schizaeaopsis macrophylla

LOCATION Virginia, USA.
AGE Early Cretaceous, Aptian (125–113 million years ago).
CHARACTERISTICS The main support structure of the leaf branched dichotomously at its base. Branches continued to divide with a strip of light-gathering tissue on either side. Strips had a central vein with smaller veins extending from it. These smaller veins also branched dichotomously. Sporangia in two rows on lobes at the tips of the leaves. Leaves bearing sporangia had scalloped edges around the veins and were rolled under to partially enclose the sporangia. Spores had a characteristic peace sign marking and distinct parallel ridges.

FOUND IN Silty sandstone.
HABITAT Sandy streambanks.
NOTES This species combined a mix of characteristics that today distinguish living genera. Dichotomizing, fingerlike leaves of *Schizaeaopsis macrophylla* linked it with living *Schizaea* and *Actinostachys*. But the Cretaceous fern had leaf tissue folded around sporangia as seen in *Schizaea* and many rows of sporangia as found in living *Actinostachys*. The spores of *Schizaeaopsis macrophylla* also combined features seen in separate living genera. *Schizaeaopsis macrophylla* spores were trilete, which conforms with the modern genus *Anemia*. Spores of living *Schizaea* have only a single slit opening. *Schizaeaopsis macrophylla* records a moment early in the evolution of the Schizaeales when features that would later separate into distinct lineages coincided in a single species.

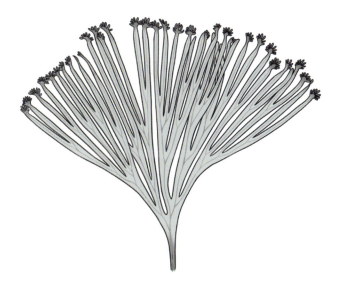

A portion of the frond of *Schizaeaopsis macrophylla*.

SALVINIALES

Since the water ferns are entirely aquatic, one might expect an excellent fossil record because these plants prefer quiet ponds where preservation is likely. But because they are small and delicate, water ferns are known mostly from their spores. Water ferns produce two sizes of spores: egg-producing gametophytes germinate from the larger spores, while sperm-producing gametophytes emerge from the smaller ones.

MORPHOLOGICAL FEATURES Modern water ferns are free-floating freshwater plants with thin, simple stems and no roots. Leaves develop in groups of three along the length of the stem. Two floating leaves are tongue-shaped with a single vein. Their upper surface is covered with water-repelling hairs that keep the surface dry and allow gas exchange for photosynthesis. The third leaf is submerged and extensively branched. Both large and small sporangia develop on the submerged leaf. However, these highly derived forms did not appear until the Cenozoic. Cretaceous species were amphibious and most still had roots, suggesting that they had not yet adopted the free-floating lifestyle.

Hydropteris pinnata

LOCATION Alberta, Canada.
AGE Late Cretaceous, Maastrichtian (72–66 million years ago).
CHARACTERISTICS Branching stems 1–2 mm (0.04–0.08 in) in diameter with simple roots. Leaves up to 6 cm (2.4 in) long with 7 to 15 elliptical lobes arranged along a central branch. Both small and large sporangia were produced in a specialized structure attached to the stem at the base of a leaf. Small spores dispersed in clusters. Large spores of the *Parazolla*–type with attached floats found within the sporangia, suggesting that the plant completed its life cycle in water, although it still retained small roots.
FOUND IN Gray siltstone.
HABITAT Pond. The combination of floating spores and roots suggests that the plant may have anchored itself at the edge of quiet ponds.
NOTES *Parazolla* spores have also been described from the Late Cretaceous of Korea and Montana, USA, indicating a wide geographic distribution for this group.

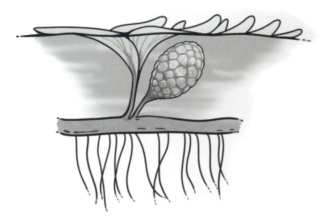

Floating leaf of *Hydropteris pinnata* and attached sporangium.

CYATHEALES

Tree ferns of this group date to the Jurassic. Tree ferns of the *Psaronius* type, members of the Marattiales, were abundant in the swamps and riverbanks of the Carboniferous but most did not survive the Late Carboniferous plant extinction. The tree habit evolved again in the Cyatheales during the Mesozoic.

MORPHOLOGICAL FEATURES Tall, straight stems with tracheids arranged in a network of strands that divide and reunite along the stem. Tracheid bundles destined for leaves branch from the main cylinder in the trunk and grow through the protective outer layers of the stem until they emerge into leaves. Tree ferns do not make wood but strengthen their trunks with reinforced cells, a swarm of roots that arise from the stem and encase it, and sturdy leaf bases. Large, scaly leaves with petioles that allow understory individuals to stretch for the canopy. Leaves three- to four-times pinnate with sporangia produced on the underside of the leaves. Sporangia develop in cup-shaped structures attached to a vein.

Alienopteris livingstonensis

LOCATION Antarctica.
AGE Early Cretaceous, Aptian (125–113 million years ago).
CHARACTERISTICS Stem 7–13 cm (2.8–5.1 in) wide. Water-conducting cells arranged in a discontinuous cylinder with complex strands of tracheids that branched and reunited along the stem. Some of the individual tracheid bundles were Y-shaped, a feature that distinguishes this Cretaceous fossil from all living species. Within the vertical stem, the water-conducting cylinder was enclosed in two layers of reinforced tissue, and then the whole stem was wrapped in sturdy leaf bases and a thick mantle of stem-based roots for more support. Water-conducting strands associated with the leaves also reinforced to support large leaves.
FOUND IN Volcanic ash of the Cerro Negro Formation.
HABITAT Tree fern and conifer forest with a diverse fern understory.
NOTES No leaves have yet been found attached to *Alienopteris livingstonensis*. However, fronds assigned the name *Lophosoria capulata* occurred in the same rock layers, suggesting that they might belong to this species. These leaves had a variety of features that united them with the Cyatheales. *Lophosoria capulata* produced *Cyatheacidites* spores, which resemble the spores of Cyatheales, both living and fossil.

Cyathocaulis yezopteroides

LOCATION Hokkaido, Japan.
AGE Late Cretaceous (100–66 million years ago).
CHARACTERISTICS Tall, straight stem about 14 cm (5.5 in) in diameter. Complex network of tracheids surrounded by a ring of sturdy, reinforced tissue. Tracheid bundles diverged from the central part of the stem and grew at a gentle angle toward the edge of the trunk and through a mantle of stem-derived roots, where they entered the petiole. These matched bundles in the leaf stem of the frond *Yezopteris polycycloides*, which supported the conclusion that these two "morphospecies" represented parts of the same plant.
FOUND IN Calcareous nodules in shallow marine shales deposited in an ancient lagoon.
HABITAT Coastal forest that also included conifers.
NOTES Today, Cyatheales tree ferns begin life in the shade under the forest canopy and grow slowly. When a gap opens, they refresh their leaves and grow rapidly toward the sunlight. This strategy likely evolved during the transition from relatively open conifer canopies typical of the Early Cretaceous and the closed angiosperm canopies of the Late Cretaceous.

Tempskya jonesii

Tympanophora simplex.

Tempskya jonesii.

LOCATION Castle Dale, Utah, USA.
AGE Mid-Cretaceous, Albian to Cenomanian (113–94 million years ago).
CHARACTERISTICS *Tempskya* produced a so-called false trunk in which 270 individual stems, leaf stems, and stem-derived roots intertwined to form an upright "trunk" about 30 cm (11.8 in) in diameter and up to 6 m (19.7 ft) tall. Individual stems branched dichotomously within the mass of roots and grew through older stems within the false trunk. This growth habit is so far unique to *Tempskya* and unknown in any living fern. A layer of reinforced cells surrounded each stem and leaf stem. Petioles were short and commonly penetrated by younger roots. Leaves emerged haphazardly from the upright stem. Leaves themselves were not preserved but the internal structure suggested that they were small and tuned to grow in full sunlight.
FOUND IN Sandstones of the Cedar Mountain, Burro Canyon, and Dakota Formations. Preserved in three-dimensional detail when mineral-rich water penetrated the stem.

HABITAT Open woodland.
NOTES *Tempskya jonesii* fossils also preserved insect burrows and fungal hyphae, suggesting that these were long-lived communities that provided habitat for other species. One *Tempskya* trunk from the Aspen Shale in Wyoming (105–100 million years ago) preserved roots of another fern from the family Lindsaeaceae within the Polypodiales, suggesting that other plants grew on *Tempskya* trunks.

Tympanophora simplex

LOCATION Northeastern Washington State, USA.
AGE Mid-Cretaceous, Albian to Cenomanian (113–94 million years ago).
CHARACTERISTICS Only a portion of the frond has been discovered, but fragments suggest it was at least 10 cm (3.9 in) wide and several tens of centimeters long. Petiole was sturdy, up to 6 mm (0.2 in) wide. Leaves were three-times pinnate. Pinnules were 1–2 cm (0.4–0.8 in) long and 2–4 mm (0.08–0.15 in) wide and arranged in an opposite pattern at about 45° from the supporting structure that bore them. Distinctive kidney bean–shaped structures on small stalks held sporangia. Up to 18 of these structures developed on each fertile pinnule.
FOUND IN Winthrop Formation, shallow marine shales, and floodplain deposits with fossil soils.
HABITAT Coastal streamside forest.

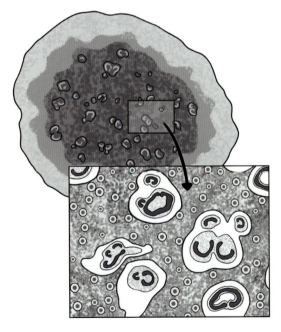

Cross section of *Tempskya jonesii* showing the bundle of roots through which slender stems meandered.

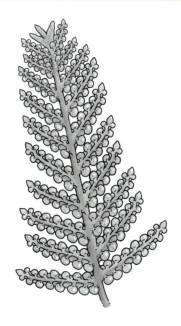

NOTES This species was also found in the Jurassic of Yorkshire, England, illustrating that the species was widespread in both space and time. *Tympanophora simplex* last appeared in the Cenomanian Frontier Formation of southern Montana, northern Utah, and western Wyoming, USA. It was among the many ferns that became extinct as flowering plants marched across the midlatitude landscape.

Specialized frond of *Tympanophora simplex* that supported sporangia.

POLYPODIALES

This most diverse group of modern ferns seems to have originated and diversified alongside flowering plants in the Cretaceous. Although petrified roots found intertwined with the stem of Albian-age *Tempskya* hint that the group arose earlier in the Mesozoic, they have a meager fossil record before the mid-Cretaceous.

Polypodiales ferns have two special ecological skills: First, they share a photoreceptor with the tree ferns (Cyatheales) that allows them to efficiently use red light wavelengths for photosynthesis. This gives the Polypodiales the ability to grow well in the deep shade that starves many plants. Second, they open their stomata quickly when a dapple of sunlight hits the leaf. This innovation maximizes photosynthesis in the darker forest understories beneath flowering plants. Quick response to sun flecks may have given the Polypodiales, along with the tree ferns, an evolutionary advantage over the groups of Mesozoic plants that were shrinking before the angiosperm juggernaut.

MORPHOLOGICAL FEATURES The Polypodiales is a diverse and highly variable group that includes about 80% of the living ferns. Two features unite them: Sporangia develop on stalks that are only one to three cells thick, and a row of thick-walled cells encircle the sporangia. These cells contract as the sporangium dries and pulls the sporangium apart to release the spores within.

Aspidistes thomasii

LOCATION Yorkshire, England.
AGE Early to Middle Jurassic (182–164 million years ago).
CHARACTERISTICS Large leaves known only from fragments. Leaves branched multiple times, at least three-times pinnate. The smallest pinnules were 7 mm (0.3 in) long and about 3 mm (0.1 in) wide. Pinnules had a single, central vein from which additional branches arose. End of the pinnule was rounded. Up to eight clusters of sporangia attached to the undersides of some pinnules. Lobes bearing sporangia generally had their edges curved under, making them

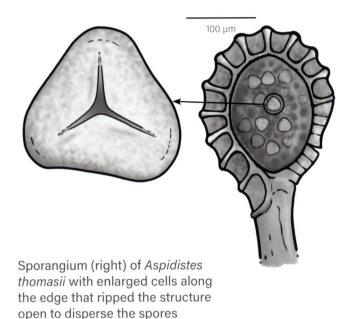

100 μm

Sporangium (right) of *Aspidistes thomasii* with enlarged cells along the edge that ripped the structure open to disperse the spores (left) inside.

Aspidistes thomasii.

appear shorter and narrower than those without sporangia. A round, protective flap of tissue attached to the central vein covered the spore cluster. The leaf surface had unicellular glands very similar to living species in the genus *Coryphopteris*.

HABITAT Streamside lowlands.

FOUND IN Clays of the Cloughton Formation.

NOTES *Aspidistes thomasii* is among the oldest fossil ferns confidently assigned to the Polypodiales and predates the group's explosive diversification. *Aspidistes thomasii* grew in a wide range of habitats and persisted through most of the Jurassic of what is today northern England and Scotland.

Athyrium asymmetricum

LOCATION Liaoning Province, China.

AGE Early Cretaceous (145–100 million years ago).

CHARACTERISTICS Large, twice-pinnate frond up to 50 cm (19.7 in) in length. Leaf was blade-shaped with a long leaf stem. The pinnules were elongated with an extended slender tip. Pinnule edges were serrated. A central vein extended into each pinnule and smaller veins branched from it. Hoof-shaped clusters of sporangia developed on the underside of the leaf in pairs on either side of the veins, about halfway to

the margin. Each cluster included globular sporangia that sported 13 thickened cells that ran across the top of the sporangium. Each sporangium produced 64 smooth spores, about 10 μm in their longest dimension.

FOUND IN Xiaoming'anbei Formation sandstones and coal.

HABITAT Wetland margins.

Athyrium asymmetricum.

NOTES The pointy tips of the frond suggest that these ferns grew in a particularly rainy climate. This shape helped rainwater drain off the surface of a leaf so that it did not flood the pores and prevent gas exchange.

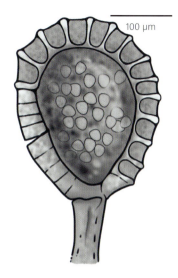

100 µm

Sporangium of *Athyrium asymmetricum.*

Astralopteris coloradica

LOCATION Utah and Colorado, USA.
AGE Early Cretaceous, Albian (113–100 million years ago).
CHARACTERISTICS Large, once-pinnate leaves with leathery photosynthetic surfaces arranged in an alternate pattern on either side of the leaf's sturdy central stem. Pinnules about 10 times longer than they were wide, with a long, pointed tip and rounded base, where they attached to the central support structure of the leaf. A single, prominent vein extended to the tip of each pinnule. Smaller veins extended from this main vein at right angles and forked two to three

times. The smallest veins formed a network to cover the entire leaf surface, an innovation that may have increased the efficiency of photosynthesis. Circular clumps of sporangia developed in pairs on either side of the pinnule main vein between the second-order veins.
FOUND IN Dakota Sandstone Formation.
HABITAT Coastal floodplain forest in a wet climate.
NOTES In 1950, Roland Brown described *Bolbitis coloradica* from the Dakota Formation in Colorado. Brown's material lacked sporangia and, to his eye, the frond structure and venation resembled modern *Bolbitis* from Southeast Asia. In 1964, James Jensen and Don Tidwell returned to Brown's sites and discovered fertile frond fragments. The sporangia were clearly different from living *Bolbitis*, leading the paleobotanists to create the new genus *Astralopteris* for Brown's original material.

Cystodium sorbifolioides

LOCATION Kachin State, Myanmar.
AGE Mid-Cretaceous, Albian to Cenomanian (113–94 million years ago).
CHARACTERISTICS Described from frond fragments preserved in amber. Pinnules had smooth margins or blunt teeth. Veins arose near the base of the pinnule and forked near the margin. Sporangia produced in circular clumps at the tip or margin of fertile pinnules. Leaves smooth. Sporangia with the typical thickened stripe of cells produced smooth spores with a peace sign mark.
FOUND IN Amber, the fossilized resin of many plants but mainly conifers and tree ferns.
HABITAT Moist forest with a conifer and tree fern overstory.
NOTES These spectacular fossils in amber also preserved fungi living on the leaf surface and evidence of insect damage to the edges of the leaf. The rich diversity of ferns and other organisms in the Myanmar

Astralopteris coloradica.

A tiny frond of *Athyrium asymmetricum* preserved in amber.

amber showed that the region's biological diversity is ancient—more than 100 million years old—and worthy of conservation.

Midlandia nishidae

LOCATION Alberta, Canada.
AGE Late Cretaceous, Campanian (84–72 million years ago).
CHARACTERISTICS This species is known from a stem with attached leaf bases arranged in a spiral.

The rhizome was about 1 cm (0.4 in) in diameter with about eight networked tracheid bundles extending along its length. Vascular tissue (phloem) that carried sugars away from the leaf surrounded each strand. The stem was sheathed in a layer of fibrous strands for support. Leaves were supplied with water by a collection of tracheid bundles that included two seahorse-shaped bundles and one to four smaller, oval bundles in the center of the leaf base.
FOUND IN Horseshoe Canyon Formation, in ironstone nodules.
HABITAT Floodplain forest.
NOTES The seahorse-shaped vascular bundles in the petiole united this fossil with living ferns in the family Blechnaceae, making it the oldest representative of this family yet discovered.

Protorrhipis choffatii

LOCATION Cercal, Portugal.
AGE Early Cretaceous, Aptian to Albian (125–100 million years ago).
CHARACTERISTICS Small (2 cm / 0.8 in long by 1.5 cm / 0.6 in wide), semicircular leaves. A straight,

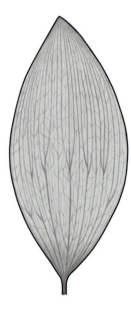

Pinna of *Protorrhipis choffatii* showing the details of venation.

relatively thick (4 mm/0.15 in) petiole attached to the flat part of the semicircle. In small (possibly immature) leaves, the margin was smooth. In larger, perhaps mature, specimens, the margin of the leaf had up to 15 pointy teeth. Primary veins were arranged like the ribs of a hand fan, radiating from the leaf stem and ending in the spaces between teeth. A narrower set of secondary veins diverged from the primary veins and extended to the leaf margin, ending in the tips of the teeth. A still finer set of veins crossed between secondary veins to form a network. A circular gland or perhaps a resting bud occurred in the space between teeth.

FOUND IN Clay of the Lusitanian Basin.

HABITAT Quiet pond or lake.

NOTES These tiny leaves were originally linked with the water lilies—they were mistakenly interpreted as flowering plants because of their pad-like shape and net-veined leaves. However, new specimens revealed that although *Protorrhipis choffatii* had net-veined leaves, it lacked the tiny veins that commonly bring water to the middle of the vein network in flowering plants. Moreover, tiny indentations scattered across the surface of the leaf indicated fern sporangia.

Wessiea oroszii

LOCATION Alberta, Canada.

AGE Early Cretaceous, Campanian (84–72 million years ago).

CHARACTERISTICS Upright stem was about 7 mm (0.3 in) in diameter with about five to seven intertwined tracheid bundles surrounded by sturdy, spherical cells that provided support and protection. Stem sheathed in leaf bases with characteristic seahorse-shaped vascular bundles and a mantle of stem-derived roots.

FOUND IN Ironstone nodules in the Horseshoe Canyon Formation.

HABITAT Floodplain forest.

NOTES *Wessiea yakimaensis* was described from Miocene-age (23 to 5 million years ago) rocks of Washington State, USA. The very similar *Wessiea oroszii* from the Cretaceous further extends the fossil record of this genus.

Winthropteris ovalis

LOCATION Northeastern Washington State, USA.

AGE Mid-Cretaceous, Albian to Cenomanian (113–94 million years ago).

CHARACTERISTICS Leaf two- to three-times pinnate. Pinnules were oval or oblong, with some having a scalloped edge. A strong, central vein entered each pinnule with smaller veins diverging in pairs and forking several times before reaching the leaf margin. Sporangia in long clusters parallel to the central vein of each fertile lobe.

FOUND IN Winthrop Formation, shallow marine shales, and floodplain deposits with fossil soils.

HABITAT Streamside forest.

NOTES Although the arrangement of the sporangia places *Winthropteris ovalis* in the modern fern family Blechnaceae, several features of the Cretaceous leaf differ from living members. For example, most living members are once-pinnate.

Winthropteris ovalis.

Antarcticycas schopfii (see p.122).

CYCADS

The earliest cycad, *Archaeocycas*, appeared in the Early Permian (about 280 million years ago). Cycads survived the end-Permian extinction and became important parts of the Mesozoic forest understory and fern savannas until they were largely displaced by fast-growing flowering plants in the Cretaceous. Despite being relatively rare today, about 300 species of cycads still inhabit tropical and subtropical areas, thriving in wet and semiarid climates from sea level to mountaintops. Although most cycads thrive in tropical warmth, some, such as *Cycas revoluta*, can tolerate below-freezing temperatures for short periods. The wide geographic range and broad ecological tolerance of cycads bear witness to an ancient lineage of tough survivors.

CYCADALES

Cycads are seed plants in which the egg cell is housed in a structure called an ovule that is fed and cared for by the parent plant. The ovule has an outer layer enclosing tissue that supports the egg cell within. An opening in one end of the ovule admits pollen, which contains sperm that fertilizes the egg cell. Upon fertilization, an embryo develops, and the ovule matures into a seed. In cycads, ovules form on short, sturdy leaves (like living *Cycas revoluta*) or in cone-like structures (like living *Encephalartos horridus*) where the supporting leaves have been modified into scales. Individual cycad plants produce either seeds or pollen. In most cycads, pollen travels from male to female cycad with the help of insects like beetles. To reward pollinators, cycads pack seed cones with starch, and many cycads produce a bit of sweet liquid at just the point where pollen must enter the ovule to fertilize the egg cell.

MORPHOLOGICAL FEATURES Cycads have stout stems guarded by an array of woody leaf bases. Stems generally do not branch, although buds in the stem may become active if the growing tip of the plant suffers damage. Some species can grow up to 18 m (59 ft) tall, but most are smaller and some keep their stem safely tucked below ground, with only a shock of tough leaves exposed to the world. A whorl of large, stiff leaves that are either entire or once-pinnate crowns each of the plant's growing tips. Inside the cycad stem, tracheid bundles intertwine with supporting tissue and canals that produce sticky resin for protection against microbial and herbivore attack. Mature cycads produce a small amount of spongy wood that helps reinforce their stems. Cycads have modest root systems that contain symbiotic cyanobacteria and fungi. Cyanobacteria in cycad roots produce a neurotoxin that can spread throughout the plant body and provide, along with compounds made by the plant, an extra line of chemical defense against herbivores. Cycads are generally slow growing and long-lived, with individuals able to live for more than a millennium.

Androstrobus prisma and *Pseudoctenis lanei*

LOCATION Hasty Bank and Gristhorpe, Yorkshire, England.

AGE Middle Jurassic, Bajocian to Callovian (170–164 million years ago).

CHARACTERISTICS Leaves assigned to *Pseudoctenis lanei* were up to 1 m (3.3 ft) long and 30 cm (11.8 in) wide. Leaf once-pinnate, with a distinctly narrower attachment of pinnae to the central rib. Pinnules ranged from 1 cm (0.4 in) wide in the lower part of the leaf to 4 mm (0.15 in) wide close to its tip. Adjacent pinnules separated by about their width. *Pseudoctenis lanei* leaves were commonly found with fragments of *Androstrobus prisma* pollen cones and disarticulated scales, the highly modified leaves that made up the cone. Individual scales were rhomboidal, about 2 x 1 cm (0.8 x 0.4 in), and covered with prism-shaped pollen sacs.

FOUND IN Floodplain sediments of the Saltwick and Cloughton Formations.

HABITAT Floodplain forest.

NOTES The stem and seed cones of this species have not yet been found. In both living and fossil cycads, seed cones are less common than pollen cones.

Antarcticycas schopfii, Delemaya spinulosa, and *Yelchophyllum omegapetiolaris*

LOCATION Fremouw Peak locality, Queen Alexandra Range, Transantarctic Mountains, Antarctica.

AGE Middle Triassic (247–242 million years ago).

CHARACTERISTICS Although they have not yet been found physically connected, the stem *Antarcticycas schopfii*, leaves *Yelchophyllum omegapetiolaris*, and pollen cones *Delemaya spinulosa* have all been found together, strongly suggesting they were produced by the same plant. Moreover, the omega-shaped arrangement of tracheids observed at the points of leaf attachment in *Antarcticycas schopfii* and in the leaf stem of *Yelchophyllum omegapetiolaris* further strengthen the connection (see illustration, p.120).

The *Antarcticycas schopfii* stem was circular in cross section and about 4 cm (1.6 in) in diameter. The largest individual discovered so far was about 13 cm (5.1 in) high. The outermost region of the stem was reinforced with sturdy fibers and filled with canals of sticky resin. Tracheids occurred in many bundles scattered throughout the stem. Each bundle contained a small amount of spongy wood. Small roots emerged from the stem. Unlike many cycad stems, the sturdy bases of leaves were not attached to the stem, suggesting that the stem remained buried in soil. Some individuals branched by dividing their growing tip in two parts, each of which produced a stem with leaves. Other specimens had small side branches that developed from buds in the stem, activated when the growing tip was damaged. *Yelchophyllum omegapetiolaris* leaves were once-pinnate with thin pinnules supplied by evenly spaced, parallel veins. Tracheid bundles entering each leaf had a characteristic inverted omega shape typical of both fossil and living cycads. *Delemaya spinulosa*

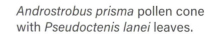

Androstrobus prisma pollen cone with *Pseudoctenis lanei* leaves.

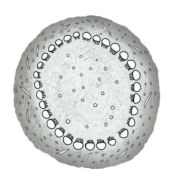

Stem cross section of
Antarcticycas schopfii.

pollen cones were about 3 cm (1.2 in) long with pollen
sacs supported by wedge-shaped scales arranged in a
tight spiral around a central stem. Each pollen sac
supporting scale had as many as five sharp spikes, which
gave the species its name. Up to eight pollen sacs
developed on each wedge. Pollen was smooth with a
single slit.

FOUND IN Mineral-impregnated peat of the Fremouw
Formation.

HABITAT Coastal swamp in a warm, temperate
polar environment in which the plant would have
experienced darkness for up to six months of the year.

Delemaya spinulosa, the pollen cone of *Antarcticycas
schopfii.*

There is no convincing evidence that *Antarcticycas
schopfii* dropped its leaves during the dark season.
Instead, it may have simply slowed growth and relied
on stored starch during the dormant period.

NOTES Two different types of fungi inhabited the roots
of *Antarcticycas schopfii.* One built saclike structures
within root tissue, while another type built a branching
network within the roots to house fungal cells. These
fossil fungi are similar to those that live in the roots of
modern cycads and provide the plant with phosphorus
and other nutrients in exchange for sugary food. These
fossils demonstrate the antiquity of the symbiotic
partnership between fungi and cycads that allows them
to thrive in poor soils. At the same time, a parasitic
fungus also inhabited the stem of *Antarcticycas schopfii.*
Paleobotanists cannot tell whether the fungus preyed
on the cycad while it was alive, but modern relatives
of this fungal group do cause plant disease. *Delemaya
spinulosa* cones contained fossil feces of insects,
showing that insects entered the cones and consumed
pollen. They probably also pollinated the cycad in
exchange for the protein-rich treat.

Beania gracilis, Androstrobus manis, Nilssonia compta, and *Deltolepis crepidota*

LOCATION Gristhorpe Bay and Hayburn Wyke,
Yorkshire, England.

AGE Middle Jurassic, Bajocian to Callovian (170–164
million years ago).

CHARACTERISTICS *Beania gracilis,* the seed cone, was
longer than 10 cm (3.9 in), with a slender, central axis
that supported a loose spiral of short stalks capped
with a small, oval scale. Two seeds, each about 1 cm
(0.4 in) in diameter, developed on each scale. The
pollen cone, *Androstrobus manis,* was about 5 cm
(2 in) long and 2 cm (0.8 in) wide, with a compact
spiral of scales each supporting several pollen sacs.
Androstrobus manis is commonly found with *Nilssonia
compta* leaves. *Nilssonia compta* were 50 cm (19.7 in)
long and 4 cm (1.6 in) wide, with a strong, central rib
that supported more or less rectangular segments of
unequal width. *Deltolepis crepidota* was the name given
to spiked stem scales that were about 1.5 cm (0.6 in)

Cycad with seed cone *Beania gracilis* (top) and the same species with pollen cone *Androstrobus manis* (above).

wide and 1.8 cm (0.7 in) long and found with *Nilssonia compta* foliage and *Androstrobus manis* cones. Living cycads commonly sport such scales to protect the stem from curious mouths.

FOUND IN Sandstones and siltstones of the Saltwick and Cloughton Formations.

HABITAT Floodplain and coastal marsh.

NOTES Although never found attached, these fossil species commonly occur together, suggesting that in life they belonged to a single species of male and female cycads.

Beania mamayi, Androstrobus wonnacottii, and Nilssonia tenuinervis

LOCATION Cloughton Wyke, Yorkshire, England.

AGE Middle Jurassic, Bajocian to Callovian (170–164 million years ago).

CHARACTERISTICS The stem branched once or twice, with each branch topped with a whorl of *Nilssonia tenuinervis* leaves. The leaves were long (about 80 cm / 31.5 in) and narrow (less than 4 cm / 1.6 in wide), with a pointed tip and short stem that flared to attach the leaf to the stem at a wide base. A strong, central rib composed of water-conducting cells and reinforcing fibers extended for the length of the leaf. *Nilssonia tenuinervis* had a straight margin that split irregularly from margin to central rib. These splits may have been damage sustained over the leaf's long life. Closely spaced fine veins extended from the central rib to the leaf margin, arranged perpendicular to the central rib. Resin glands that produced defensive chemicals developed between adjacent pairs of veins. *Beania*

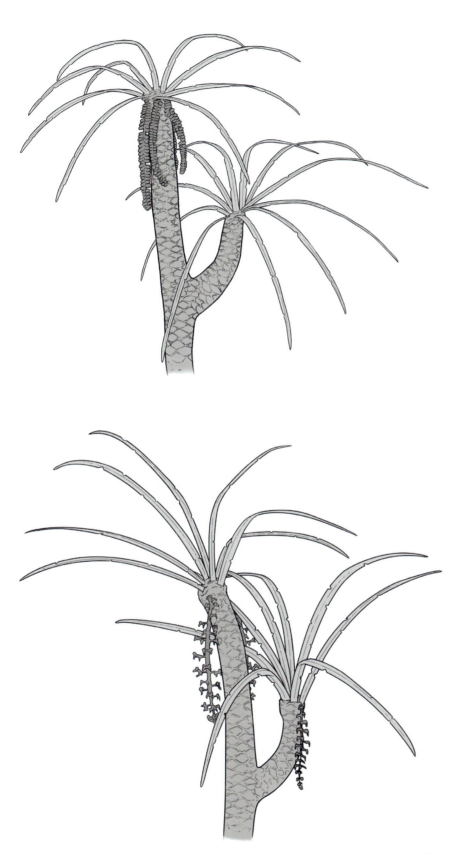

Cycad *Beania mamayi* with pollen cone (top) and seed cone (above).

mamayi, the seed-producing cone, was at least 18 cm (7.1 in) long, fibrous, infused with resin, and flexible, with the cone dangling from the plant. Cone scales were about 1.6 cm (0.6 in) long and widely spaced on the cone's main axis. Seeds were about 4 mm (0.15 in) in diameter, with a thin, fleshy layer covering the hard seed within. *Androstrobus wonnacottii*, the pollen cone, was 3.5–7.5 cm (1.4–3 in) long and about 1 cm (0.4 in) wide, with a rounded top and a base that connected directly to the stem at the base of a leaf. Tough, resin-infused cone scales spiraled loosely around a central core. A cluster of cylindrical pollen sacs hung from each cone scale.

FOUND IN Floodplain ponds and shallow marine sands.

HABITAT Vegetated floodplain.

NOTES Attentive readers will note that the foliage genus *Nilssonia* is spelled *Nilsonia* by some authors. Adolphe-Théodore Brongniart created the name *Nilsonia* in 1825 for cycad-like foliage from the Late Triassic of Sweden. Brongniart intended the name to honor Sven Nilsson but misspelled the family name in the original publication. Heinrich Bronn amended the name to *Nilssonia* in 1834 and that spelling was adopted by Tom Harris when he reported 10 species from the Jurassic of Yorkshire. A strict interpretation of botanical naming would favor the older spelling. We follow Harris in honoring Brongniart's original intent with the double *s*.

Paleobotanist Tom Harris united these three form species into a single whole plant based on their co-occurrence in the Cloughton Wyke locality. He supported his argument with details of the numerous resin glands that dot the plant's surface.

Bjuvia simplex and Palaeocycas integer

LOCATION Southern Sweden.
AGE Late Triassic (209–201 million years ago).
CHARACTERISTICS *Palaeocycas integer* was a spoon-shaped, highly modified leaf that supported developing seeds. The bowl of the spoon was about 2 cm (0.8 in) across on a short stem. Two pairs of ovules grew on either side of the spoon's handle, which emerged in a whorl from the cycad's tip in place of photosynthetic leaves. *Bjuvia simplex* leaves were large, perhaps larger than 1 m (3.3 ft) in length and 20 cm (7.9 in) wide, with entire margins and a strong, central rib. Closely spaced fine veins emerged perpendicular to the central rib and extended straight to the edge of the leaf. Some veins near the base of the leaf divided into two to maintain even vein spacing. Pollen cones have not been identified.
FOUND IN Clay layers between coal seams of the Bjuv Member of the Höganäs Formation.
HABITAT Swamp forest.
NOTES Leaves that resemble *Bjuvia simplex* (called *Taeniopteris*) have been reported from the Permian and throughout the Mesozoic. *Taeniopteris* has been linked with a variety of plants, including cycads, marattialean ferns, and extinct seed plants in the Bennettitales.

1 cm

Portion of the cone holding seeds of *Palaeocycas integer*.

However, details of the stomata of *Bjuvia simplex* link this leaf with the cycads, and its common association with *Palaeocycas integer* unites the forms into a single species.

Leptocycas gracilis and Pseudoctenis foliage

LOCATION Chatham County, North Carolina, USA.
AGE Late Triassic (237–227 million years ago).

Leptocycas gracilis.

CHARACTERISTICS *Leptocycas gracilis* was a slender stem 3–5 cm (1.2–2 in) wide, with a crown of *Pseudoctenis*–type leaves. Leaves were up to 30 cm (11.8 in) long and once-pinnate, with a long petiole that attached to the plant with a wide base that persisted on the stem after the leaf died. Pinnules were about 4.5 cm (1.7 in) long and 3.5 cm (1.4 in) wide, with fine, straight veins. Pinnules were arranged randomly along the leaf's central axis with some pinnae opposite and some alternate. Small, scale-like leaves surrounded the growing tip of the stem, probably protecting developing leaves within. Some of these scale leaves remained attached to the stem after the leaves had unrolled. Seeds and pollen cones are not known.

FOUND IN Boren Clay Pits of the Pekin Formation in the Deep River Basin.

HABITAT Lake margin or floodplain forest with a conifer overstory and fern understory and horsetails growing along sandy stream margins. Forests developed in valleys that opened as Pangaea began to break apart in the Late Triassic.

NOTES A variety of other cycads are the most abundant plant fossils at this locality. The rift valley wetlands also supported fish and crayfish-like animals that burrowed into riverbank mud. Crocodile-like reptiles basked, and large, bipedal dinosaurs roamed the banks.

Mature *Ginkgo* (see p.131).

GINKGOS AND RELATIVES

Ginkgo biloba is the last living species of this ancient group. And like humans themselves, it is the last survivor of a long and diverse lineage.

CZEKANOWSKIALES

This group takes its name from fringe-like foliage called *Czekanowskia*. Like the foliage of ginkgos, *Czekanowskia* leaves attached in bundles to short shoots. Seed-bearing structures found in the Late Permian of Italy may be the oldest members of the Czekanowskiales, but their heyday was the Triassic, Jurassic, and Early Cretaceous, when they inhabited forests throughout the Northern Hemisphere. Paleobotanists presume that the Czekanowskiales were trees like their ginkgo relatives or perhaps woody shrubs, but no wood has been conclusively linked to the distinctive foliage. Czekanowskiales grew in many middle- and high-latitude forests where rainfall was plentiful but only dominated in the area that is today the Amur region of Siberia. In East Greenland, thick leaf mats on the forest floor suggest that Czekanowskiales may have shed their foliage seasonally. Like many Triassic and Jurassic seed plants, the Czekanowskiales slowly faded from the ecological stage with the rise of flowering plants, and by the end of the Cenomanian, they were extinct.

MORPHOLOGICAL FEATURES Long, narrow, fringe-like leaves were borne in clumps on compact, woody shoots that were surrounded by smaller, stiff, scalelike leaves. Seeds developed on long, dangling structures that displayed a series of clamshell-like pods. Several ovules were tucked inside the enrolled edge of each valve of the pod. Pollen organs and wood are not known.

Czekanowskia chinensis

LOCATION Shiguai Coal Mine, Shiguai Basin, Inner Mongolia, China.
AGE Middle Jurassic, Aalenian to Bajocian (174–166 million years ago).
CHARACTERISTICS Six long (about 10 cm / 3.9 in), thin (less than 1 mm / 0.04 in wide) leaves attached in a bundle on a short shoot. One vein entered each leaf at its base and divided with the blade of the leaf. Each of the six leaves widened to about 2 mm (0.08 in) before dividing two or three times to produce a fringe. Stomata

Feathery foliage of *Czekanowskia chinensis*.

appeared in rows on both the upper and lower surface of the leaf.

FOUND IN Fine-grained sandstones and siltstones from ponds of the Zhaogou Formation.

HABITAT Floodplain forest in which *Czekanowskia chinensis* joined ginkgos and conifers in the canopy, with an understory of ferns and cycads.

NOTES No wood, seeds, or pollen organs have been found in association with *Czekanowskia chinensis* leaves. Younger deposits in Inner Mongolia have yielded pollen-bearing structures called *Ixostrobus*, which may be associated with *Czekanowskia* leaves. However, the pollen cones were flimsy, fell apart during pollen dispersal, and preserved poorly, so little is known about them. Moreover, some species of *Ixostrobus* have been linked with cycads, which calls into question their connection to *Czekanowskia*.

Solenites vimineus and *Leptostrobus cancer*

LOCATION Gristhorpe Plant Bed, Yorkshire, England.
AGE Middle Jurassic, Bajocian to Callovian (170–164 million years ago).
CHARACTERISTICS *Solenites vimineus* leaves were 15–20 cm (5.9–7.9 in) long and about 1 mm

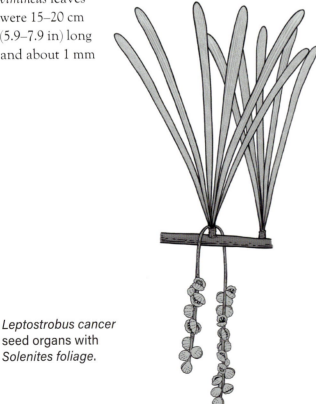

Leptostrobus cancer seed organs with *Solenites* foliage.

(0.04 in) wide. A single vein entered each leaf and ran, unbranched, to its tip. Leaves were displayed in a bunch of 10 to 15 on short (5 x 5 mm / 0.2 x 0.2 in) shoots that were covered in robust scales. The short shoot, scales, and leaves were shed as a unit. The tips of the leaves were pointed to help shed water in rainy climates. The seed organ, *Leptostrobus cancer*, was consistently found with *Solenites vimineus*, suggesting that they were produced by the same tree. *Leptostrobus cancer* had a dangling stem on which clamshell-like pods developed. Each pod was rounded, about 5 mm (0.2 in) in diameter, and arranged at 5 mm (0.2 in) intervals on the slender stem. The edges of each valve curled in to support up to five developing seeds within each of the two valves.

FOUND IN Silty sandstones of the Gristhorpe Member of the Cloughton Formation.

HABITAT Floodplain forest.

NOTES *Leptostrobus cancer* has also been reported from the Late Triassic Yangcaogou Formation and the Early Cretaceous Yixian Formation of Liaoning Province in China, demonstrating that plants sporting this reproductive structure changed little for perhaps 100 million years before being displaced from warm temperate forests by drying climate and angiosperm invaders.

Stenopteris williamsonis

LOCATION Gristhorpe Plant Bed, Yorkshire, England.
AGE Middle Jurassic, Bajocian to Callovian (170–164 million years ago).
CHARACTERISTICS The largest examples of this leaf were about 15 cm (5.9 in) long and 6 cm (2.4 in) wide. Each leaf had a short stalk with an expanded base where it attached to the

Stenopteris williamsonis foliage.

stem of the parent plant. The central rib of the leaf supported numerous feathery side branches with a single vein entering each of the slender lobes. The leaf was not extensively armored.

FOUND IN Silty sandstones of the Gristhorpe Member of the Cloughton Formation.

HABITAT The edges of quiet floodplain ponds.

NOTES These leaves are relatively rare and were probably borne on a large tree that stood on stable ground farther from the active floodplain. The shape of the leaf base suggests that leaves may have been shed periodically, perhaps seasonally.

GINKGOALES

The heyday of ginkgo diversity began in the Late Triassic and extended through the Jurassic. Ginkgo abundance and diversity declined precipitously as flowering plants expanded their range in the mid-Cretaceous, but a few ginkgo leaves commonly appear in plant fossil assemblages from the Cretaceous and into the Cenozoic. During the Mesozoic, ginkgos had a worldwide distribution and preferred cool, humid climates in the high latitudes. Ginkgos probably evolved the tendency to shed their leaves in response to dark polar conditions rather than freezing, although the trait helped them survive the cooling Cenozoic climate. Ginkgos became extinct in the Southern Hemisphere soon after the end-Cretaceous extinction event. Both their geographic range and diversity continued to shrink during the Cenozoic as the pole became frigid and the interiors of continents dried. Today, only a single species, *Ginkgo biloba*, persists.

MORPHOLOGICAL FEATURES Ginkgos are woody canopy trees (see image on p.128). Slow growing and long-lived, they may have reached enormous diameters. Very little fossil wood specifically attributed to ginkgos has been reported. This may be because ginkgo wood resembles that of many conifers. Alternatively, the thin walls of ginkgo tracheids, and the lack of antimicrobial pitch, allowed ginkgo wood to decompose quickly. Young ginkgos grew straight, with side branches emerging at intervals. However, as the tree matured, its architecture varied considerably. Leaves of most

Young *Ginkgo* trees have a conifer-like architecture.

species were fan-shaped and could be entire (like the mature leaves of living *Ginkgo biloba*), deeply divided, or fringe-like. Leaves developed on short shoots in which internodes did not elongate. However, unlike the Czekanowskiales, short shoots persisted on ginkgos, and leaves and seeds were shed individually. Living *Ginkgo biloba* has separate seed- and pollen-producing individuals. This was likely true of its extinct relatives as well. Seeds developed on short shoots as naked ovules hanging from short stalks. After pollination, a hard, inner shell enclosed the embryo and its food reserve, with a fleshy, outer layer forming around the shell. Pollen organs also developed on short shoots. They were flimsy, with pollen sacs arrayed in whorls on a pendant stalk. Living *Ginkgo biloba* is primarily wind pollinated, and the delicate, catkin-like pollen organs of fossil ginkgos suggest that the modern tree inherited this strategy from its ancestors.

Baiera africana

LOCATION Umkomaas, KwaZulu-Natal, South Africa.
AGE Late Triassic, Carnian (237–227 million years ago).
CHARACTERISTICS Leaves were fan-shaped, approximately 8 cm (3.1 in) long and 4 cm (1.6 in)

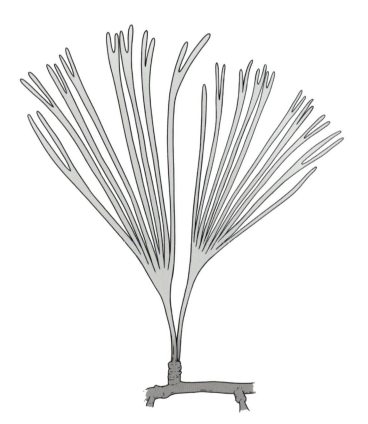

Baiera africana foliage.

wide, with two veins entering a well-defined petiole about 2.5 cm (1 in) long. The leaf divided into two lobes with one vein entering each. Veins continued to divide in each lobe at intervals of a few millimeters until they reached the margin of the leaf. About 1 mm (0.04 in) of leaf tissue on either side of each vein filled in the leaf to create the light-gathering surface. This gave the leaf a fringed appearance. *Baiera africana* is known only from dispersed leaves.
FOUND IN Mudstones of the Molteno Formation.
HABITAT Floodplain forest.
NOTES In addition to its original description from South Africa, *Baiera africana* has been found in rocks from the Late Triassic (Carnian to early Norian) of Biobío Province in Chile, the state of Rio Grande do Sul in Brazil, and the Cañadón Largo Formation of Argentina, showing that *Baiera africana* trees were widespread across the midlatitudes of Gondwana.

Ginkgo adiantoides

LOCATION Montana and North Dakota, USA.
AGE Late Cretaceous, Maastrichtian (72–66 million years ago).
CHARACTERISTICS Fan-shaped leaves were about 5 cm (2 in) long and 6 cm (2.4 in) wide, with a short petiole up to 1 cm (0.4 in) long and 4 mm (0.15 in) wide. The sides of the fan were straight, with the top curved outward and undulating. Two veins entered the leaf and forked multiple times so that they maintained nearly constant spacing of about 0.5 mm (0.2 in) as the leaf widened toward the margin.

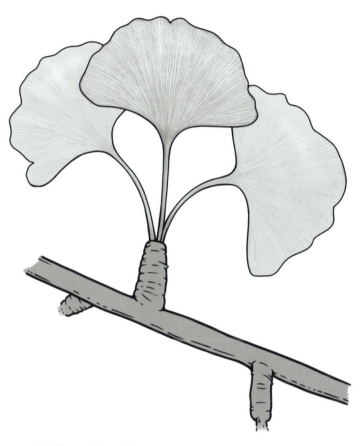

Ginkgo adiantoides.

Ginkgo apodes and Sphenobaiera-type foliage

LOCATION Liaoning Province, China.

AGE Early Cretaceous, Berriasian to Aptian (145–113 million years ago).

CHARACTERISTICS The seed-producing structure *Ginkgo apodes* developed on an approximately 3 cm (1.2 in) long stalk that widened into a nob supporting up to six ovules, each in a cuplike structure. Mature seeds were nearly spherical, about 1 cm (0.4 in) in diameter, with a thin, fleshy covering and smooth skin. *Sphenobaiera*–type leaves were attached to the parent plant without a petiole. The leaf base was about 0.5 cm (0.2 in) wide and 5 cm (2 in) long. Above this base, the leaf divided in two, with each of these segments dividing into three blades each about 0.5 cm (0.2 in) wide for an overall fan shape.

FOUND IN Sandstones and coal of the Xiaoming'anbei Formation.

Stomata only occurred on the bottom of the leaf and were commonly surrounded by short hairs, which helped conserve water.

FOUND IN Fine-grained sandstones and mudstones of the Hell Creek Formation.

HABITAT River margin and estuarian forest.

NOTES *Ginkgo adiantiodes* grew in Europe and North America from the Early Cretaceous through the Miocene (disappearing about 15 million years ago), with this species among the last to be found in these regions. *Ginkgo adiantiodes* resembles living *G. biloba* and some have suggested direct ancestry between the Mesozoic and modern species. However, evolutionary relationships should also consider the seed- and pollen-producing structures, branch form, and wood, but at present, *Ginkgo adiantiodes* is known only from its fallen leaves.

Ginkgo apodes with *Sphenobaiera foliage.*

HABITAT Wetland margins where several species of ginkgos shared the canopy with conifers and an understory of ferns.
NOTES *Ginkgo apodes* shared the forest canopy with G. *pediculata*.

Ginkgo huttonii

LOCATION Yorkshire, England.
AGE Middle Jurassic, Bajocian to Callovian (170–164 million years ago).
CHARACTERISTICS Leaves of *Ginkgo huttonii* were fan-shaped, 3–4 cm (1.2–1.6 in) across, and had a slender petiole. The leaf was divided into six segments with rounded or notched tips. Round seeds about 1 cm (0.4 in) in diameter are frequently found with these leaves, suggesting that they belong to the same parent plant. Rare pollen organs for *Ginkgo huttonii* were slender, several centimeters long, with a spiral of small scales, each of which displayed a pair of pollen sacs. Pollen was boat-shaped, smooth, and opened along a single slit.
FOUND IN Floodplain sandstones and mudstones of the Saltwick, Cloughton, and Scalby Formations.
HABITAT Floodplain forest that included *Solenites vimineus* and an understory of cycads, Bennettitales, and ferns.
NOTES *Ginkgo huttonii* is the most abundant ginkgo in the Yorkshire Jurassic and is the most common

fossil at Scalby Ness, where pollen organs and seeds have also been discovered. Leaves were also preserved at Hasty Bank, Hayburn Wyke, Gristhorpe Bay, Cloughton Wyke, and Cayton Bay, demonstrating that the species was present, if seldom abundant, throughout the Jurassic floodplain forest.

The pollen and pollen cone of *Ginkgo huttonii* resembled that of living G. *biloba*, showing that this feature was conserved over tens of millions of years. Pollen cone form also argues that *Ginkgo huttonii* was wind pollinated like living G. *biloba*.

Ginkgo pediculata and Ginkgoites manchurica

LOCATION Liaoning Province, China.
AGE Early Cretaceous, Berriasian to Aptian (145–113 million years ago).
CHARACTERISTICS The seed-producing structure *Ginkgo pediculata* developed on a short stalk with three to five branches. Each branch was topped with a small, cup-shaped structure that held an oval ovule that was about 1.5 cm (0.6 in) in longest dimension. The opening where the pollen entered the ovule faced away from the plant. In mature seeds, the fleshy, outer covering was 2–3 mm (0.8–1 in) thick and surrounded a hard shell approximately 10 mm (0.4 in) long and 9 mm (0.35 in) in diameter. Seeds at different stages of maturity were found together on the same short stalk. *Ginkgoites manchurica*

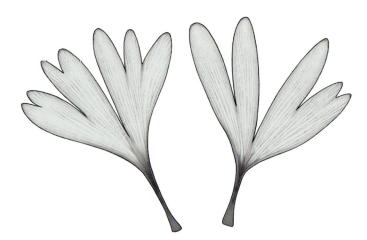

Ginkgo huttonii.

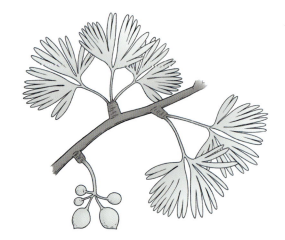

Ginkgo pediculata.

leaves were fan-shaped or nearly semicircular, with a 4.5 cm (1.8 in) stalk. The leaf was divided into two parts, each of which then further divided two to four more times into 8 to 24 segments, with four to five veins each. Stomata occurred only on the underside of the leaf in rows that extended from base to leaf margin. The connection between leaves and seeds is supported by association on the same fossil slab and by similarities in the details of cells on the surface of each.

FOUND IN Sandstones and coal of the Xiaoming'anbei Formation.

HABITAT Wetland margins where several species of ginkgos shared the canopy with conifers and an understory of ferns.

NOTES *Ginkgoites manchurica* was originally described under the name *Ginkgo manchurica* because of its resemblance to living *Ginkgo*. Some paleobotanists prefer to use the suffix *-ites* to indicate a fossil form that resembles a living form. *Ginkgoites manchurica* was one of the most common leaves from the Early Cretaceous floras of Northeastern China.

Ginkgo yimaensis

LOCATION Yima, western Henan Province, China.
AGE Middle Jurassic, Aalenian to Bajocian (174–168 million years ago).
CHARACTERISTICS Branches had long and short shoots and produced a variety of leaf forms. Leaf shapes ranged from oval leaves about 6 cm (2.4 in) long and 3 cm (1.2 in) wide with a notch at the tip, to those divided into two lobes with rounded tips, to fan-shaped leaves divided into four segments each with a rounded and deeply notched tip. Seeds developed on a slender stem about 2 cm (0.8 in) long that divided into two or three branches, each another 1 cm (0.4 in) long. At the end of each branch, a small, cup-shaped structure held a single ovule. The end of the ovule where pollen entered faced away from the plant. Mature seeds were nearly spherical and a little more than 1 cm (0.4 in) in diameter with a 2 mm (0.08 in) thick, fleshy covering. Although no pollen organs have yet been found, the seeds and leaves were covered with delicate, purse-shaped pollen grains. The abundance of pollen suggested that they were wind pollinated.

FOUND IN Siltstones below a coal seam in the Yima Formation.

HABITAT Forest at the margin of a wetland that included several other species of ginkgo and a fern understory.

NOTES Leaves and seeds were found together in the same

Ginkgo yimaensis.

Ginkgo yimaensis.

Ginkgo yimaensis had leaves in a variety of shapes and sizes, and seeds that were egg-shaped.

fossil bed. In addition, leaves and seeds shared a variety of surface features, such as the shape of cells, hairs, and glands, suggesting that they were produced by the same plant.

Glossophyllum florinii

LOCATION Niederösterreichische Nordalpen in the Northern Calcareous Alps of northeastern Austria.
AGE Late Triassic, Carnian (237–227 million years ago).
CHARACTERISTICS This species of purported ginkgo affinities was described only from foliage. Strap-shaped leaves were 10–12 cm (3.9–4.7 in) long and 1–2 cm (0.4–0.8 in) wide at the widest point. Leaves were attached to the shoot by a wide base similar to that of *Sphenobaiera*. Leaves were thick and leathery with stiff, reinforcing tissue between parallel veins. Stomata developed on both sides of the leaf, and the waxy, waterproof covering on the leaf surface was unusually thick.
FOUND IN Nearshore marine sandstones of the Lunzer Sandstein.
HABITAT Coastal or estuarian forests that also

included conifers, cycads, horsetails, ferns, and rare ginkgos.
NOTES At first glance, the slender, strap-shaped leaves of *Glossophyllum florinii* do not seem allied with the fan-shaped leaves typical of ginkgos. However, the details of their surface cells and stomata point toward shared ancestry with ginkgos. To date, no seeds or pollen organs have been found in association with these leaves. Leathery leaves with thick waterproofing help plants growing in dry climates or in salty soil to conserve water. Thick, reinforced leaves deter predators and support long-lived leaves that would lose their photosynthetic capability if tracheids kinked.

Hamshawvia balausta, Stachyopitys lacrisporangia, and Sphenobaiera foliage

LOCATION Paraná Basin, state of Rio Grande do Sul, Brazil.
AGE Late Triassic, Carnian (237–227 million years ago).
CHARACTERISTICS The seed-producing structure *Hamshawvia balausta* developed on a slender stalk

Glossophyllum florinii foliage.

Hamshawvia balausta seeds with *Sphenobaiera* foliage.

that divided in two to support a pair of flat, oval structures that held five pairs of ovules on the underside of each. At the time when the fossil was preserved, ovules were at the point of pollination and only about 1 mm (0.04 in) long. They would have swelled into a clump as they matured. The pollen-producing structure *Stachyopitys lacrisporangia* consisted of a delicate stalk several centimeters long, with 2 mm (0.08 in) side branches arising at irregular intervals. Each branch ended in a tuft of about a dozen elliptical pollen sacs, each about 2 mm (0.08 in) long. Reproductive structures were found with several species of *Sphenobaiera* foliage, including *Sphenobaiera tenuinervis* and *S. sulcata*. *Sphenobaiera tenuinervis* was fan-shaped, about 8 cm (3.1 in) long and 5 cm (2 in) wide. The leaf attached to the shoot with a 2 mm (0.08 in) wide base. Two veins entered each leaf and repeatedly divided, with about 1 mm (0.04 in) of leaf tissue on either side of each vein to give the leaf a fringed appearance. *Sphenobaiera sulcata* was smaller, only about 5 cm (2 in) long and 2 cm (0.8 in) wide, and more clearly divided into two fringed lobes.

FOUND IN Mudstones of the Passo da Tropas Member of the Santa Maria Formation.

HABITAT Floodplain and lakeside forest.

NOTES *Hamshawvia balausta, Stachyopitys lacrisporangia*, and *Sphenobaiera* foliage were found together in the same beds but never in connection. At least one other species of *Hamshawvia* was found in the same deposits, as was the leaf form *Baiera africana*. Given the tossed salad of floodplain preservation, we can only conclude that several different ginkgo species grew in the floodplain forest. Complicating the picture, *Sphenobaiera tenuinervis* and *S. sulcata* were described as separate species, although ginkgos are known to produce a variety of leaf shapes and sizes on the same tree.

Karkenia henanensis and *Sphenobaiera* foliage

LOCATION Yima, western Henan Province, China.

AGE Middle Jurassic, Aalenian to Bajocian (174–168 million years ago).

CHARACTERISTICS The seed-producing cone *Karkenia henanensis* was about 4 cm (1.6 in) long and 2 cm (0.8 in) wide with more than 100 densely packed ovules on short stalks. The stalk of each ovule was curved back toward the central stem of the cone so that the end of the ovule where pollen would enter faced the center of the cone. Individual ovules were 2–3.5 mm (0.08–0.14 in) long and 1–1.5 mm (0.04–0.06 in) wide. *Sphenobaiera* leaves found on the same rock slab were shaped like an elongated fan about 7 cm (2.8 in) long and 3 cm (1.2 in) wide. Leaves had a wide attachment to the stem and divided into four segments, each less than 1 cm (0.04 in) wide.

FOUND IN Siltstones below a coal seam in the Yima Formation.

HABITAT Forest at the margin of a wetland that included several other species of ginkgo.

NOTES In most ginkgos, just a few ovules were displayed on short stalks. In *Karkenia*, ovules were clumped together into a structure resembling a conifer-like cone. However, consistent association of *Karkenia* with ginkgo foliage shows that it belonged to the ginkgo group.

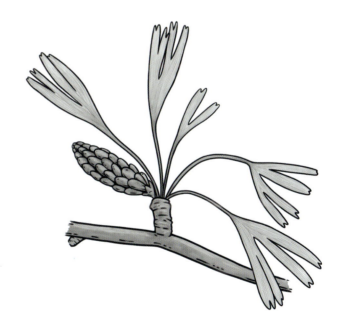

Karkenia henanensis seeds with *Sphenobaiera* foliage.

CONIFERS

Conifers dominated many Triassic and Jurassic forests and their wood constitutes one of the most common and conspicuous plant fossils of the time. Although they lost significant diversity during the Cretaceous Flowering Plant Revolution, many of their major groups survived and remained abundant in some ecosystems. Conifers have found an evolutionary recipe for success that includes simple, yet flexible, architecture, reliable reproduction, and well-honed defenses tested by more than 250 million years of predators.

Conifers are seed plants, meaning that fertilization by sperm carried within pollen takes place while the ovule is still attached to the parent plant. Moreover, the embryo begins life within a seed that is packed with food to sustain it for a period of dormancy and during early growth. Unlike cycads and flowering plants, conifers never evolved pollination relationships with animals. Instead, most conifers opt to produce vast amounts of pollen and cast it to the wind.

MORPHOLOGICAL FEATURES Conifers are mostly woody trees with a strong vertical stem from which branches emerge in whorls perpendicular to the trunk. Branches grow outward from the trunk, giving conifers grown in the open their conical shape. However, when trees grow close together, lower branches experience shade and die, leaving the mature tree with a straight, branchless trunk and a tuft of leafy branches in the canopy. Despite this common form, some lineages have evolved more irregular and bushy branching patterns. Conifer wood consists only of tracheids and is infused with sticky, chemical-laced resin that covers wounds and deters herbivores.

Foliage comes in a variety of shapes, from slender needles to millimeter-wide scales to stiff, spiked tongues and straps. Conifer leaves generally develop in spirals around the smallest branches or as scales pressed tightly to the top and bottom of the branches, with larger leaves extending from

RIGHT: Conifers come in a variety of shapes and sizes.

OPPOSITE: The "living fossil" *Metasequoia glyptostroboides* found near Moudao in Hubei Province, China.

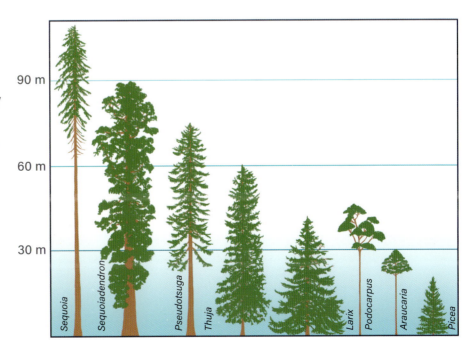

the sides. The leaves tend to be tough, with a thick outer covering that protects them from herbivore attack and water loss. Leaves must be sturdy because many conifers retain them for several years. Only a handful of conifers drop all their leaves seasonally. Conifer cones mimic the architecture of the tree: a central stem from which emerge whorls or spirals of bracts and scales. Conifer ovules develop on a papery scale resting on a woody bract. Many conifer seeds have filmy wings that aid dispersal. Pollen cones are generally small and flimsy compared to woody seed cones. In pollen cones, pollen sacs develop on cone scales.

ARAUCARIACEAE

Conifers in the Araucariaceae grew in forests worldwide during the Jurassic and Cretaceous but became extinct in the Northern Hemisphere early in the Cenozoic. Today, three genera, *Araucaria*, *Agathis*, and *Wollemia*, survive in temperate regions of South America and tropical and subtropical regions of Australia, New Zealand, Polynesia, Micronesia, Melanesia, and Indonesia.

MORPHOLOGICAL FEATURES Trees with a straight woody stem and three to seven horizontal branches emerging in whorls at regular intervals. Stems have a core of spongy cells in the center and a zone of resin-filled cells just under the bark. Leaves are arranged in a spiral on branches and may be small and hooked or broad and oval. Pollen cones are small and flimsy, with each scale bearing four to twenty pollen sacs. Seed cones are larger, woody, spherical or egg-shaped, with the cone bract and scale fused into a single structure.

Araucaria mirabilis and *Masculostrobus altoensis*

LOCATION Cerro Cuadrado Petrified Forest, Patagonia, southern Argentina.
AGE Middle to Late Jurassic, Bathonian to Oxfordian (165–156 million years ago).
CHARACTERISTICS Seed cones spherical, ranging from 3–10 cm (1.2–3.9 in) in diameter. The central axis of the cone had a spongy core, matching the wood of the branches. Woody bracts emerged from the central axis in a spiral and each housed one seed about 1 cm (0.4 in) long. The pollen cone *Masculostrobus altoensis* was found in the same deposits as abundant *Araucaria mirabilis* seed cones. Pollen cones were about 2 cm (0.8 in) long and 8 mm (0.3 in) wide, with a slender central axis and a spiral of delicate scales that supported sacs of flimsy, spherical pollen. Wood bore characteristic structures called burls that, in living conifers, protect growing cells near the base of the

Araucaria mirabilis.

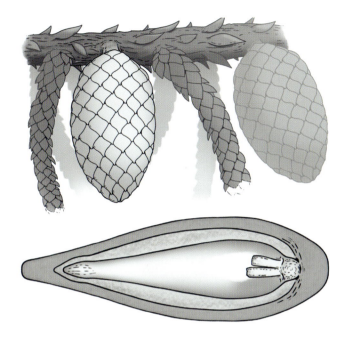

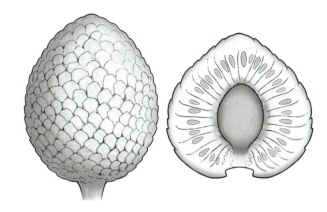

ABOVE LEFT: Seed cones of *Araucaria mirabilis*.

ABOVE RIGHT: *Araucaria mirabilis* in cross section showing seeds on individual cone scales.

LEFT: The interior of a seed showing the tiny, two-part embryo.

trunk. When the aboveground part of the tree dies, these cells can produce new shoots that allow it to regrow from its roots.

FOUND IN La Matilde Formation, which includes volcanic ash deposits associated with Cerro Cuadrado, Cerro Alto, and Cerro Madre e Hija.

HABITAT Midlatitude upland forest.

NOTES *Araucaria mirabilis* cones are large, charismatic and widely displayed in museums worldwide. These cones have been found attached to petrified wood (*Araucarioxylon*) that resembles wood of living *Araucaria*. Similar seed cones have been discovered in the Early Cretaceous of India. Although probably not the same species, these younger fossils may represent a close relative and indicate that the group was widespread across Gondwana.

Brachyphyllum mamillare and *Araucarites phillipsii*

LOCATION Yorkshire, England.

AGE Middle Jurassic, Bajocian to Callovian (170–164 million years ago).

CHARACTERISTICS *Brachyphyllum mamillare* leaves were about 1.5 mm (0.06 in) long and 2 mm (0.08 in) wide, triangular, arranged in a spiral around the branch, and thick and fleshy in life. Pollen cones about 1 cm (0.4 in)) long and 5 mm (0.2 in) wide are commonly found attached to the ends of dispersed leafy branches. Pollen cones had a central axis to which a spiral of rhomboidal scales attached. Pollen was spherical and thin-walled, with minute raised bumps. Seed cones have never been found attached to *Brachyphyllum mamillare* and therefore carry the name *Araucarites phillipsii*. The seed cones appear to have fallen apart at maturity, and so dispersed triangular bracts are most common. Bracts were about 1.5 cm (0.6 in) long and 1.3 cm (0.5 in) wide, with a sharply pointed tip. A single seed, approximately 1 cm (0.4 in) long and 5 mm (0.2 in) wide, developed on each bract.

FOUND IN Common in all fossil-bearing localities in Yorkshire.

Brachyphyllum mamillare foliage.

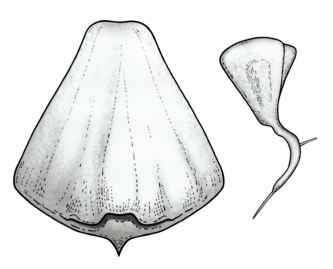

Cone scale of *Araucarites phillipsii*.

HABITAT Floodplain and coastal forest.
NOTES Fossil araucarian wood accumulated as a logjam in the ancient river at present-day Saltwick Bay near the village of Whitby. Submerged in stagnant water, the wood resisted decay and, when buried, transformed into the semiprecious material jet. Whitby jet has been carved since the Bronze Age, and during the Victorian era, black charms and lockets of Whitby jet were popular remembrances of loved ones who had passed. *Brachyphyllum* foliage has also been reported attached to twigs associated with other conifer groups.

Araucaria vulgaris

LOCATION Hokkaido, Japan.
AGE Late Cretaceous, Turonian to Santonian (94–82 million years ago).
CHARACTERISTICS Woody branches up to 10 cm (3.9 in) in diameter were found with attached foliage and a seed cone. Wood had densely packed tracheids and bark broken into

regular diamond-shaped pieces 5–7 cm (2–2.8 in) wide and 0.5–1.1 cm (0.02–0.4 in) high on the thick parts of the branch. Small branches divided at irregular intervals to form a flattened array for gathering light. Leaves were small, 2–3 mm (0.08–0.1 in) long and 6–7 mm (0.2–0.3 in) wide, rhomboidal, overlapping, and pressed tightly against the branch. The attached seed cone was spherical, about 2.2 cm (0.9 in) in diameter, and attached to the branch by a short stem. About 30 bracts attached to the cone axis. Each bract had an upturned outer tip and produced one seed.
FOUND IN Calcareous nodules in lagoon sediments of the Upper Yezo Group.
HABITAT Coastal forest.
NOTES This new whole-plant species was created when seed cones previously called *Araucaria nihongii* were found attached to branches with foliage named *Yezonia vulgaris*.

Araucaria vulgaris.

CHEIROLEPIDIACEAE

The Cheirolepidiaceae first appeared in the Ladinian age of the Middle Triassic and were common at low latitudes through the early Cretaceous. They disappeared from most regions by the Turonian but reappeared in South America during the Maastrichtian and were common in the regional flora. The group survived the end-Cretaceous extinction event and persisted into the Paleocene, finally becoming extinct at the end of the epoch. The Cheirolepidiaceae included trees and shrubs, some of which showed adaptations to water stress. This evolutionary trend may explain their disappearance in the Turonian—lowland species became extinct while survivors occupied dry habitats far from sites of fossilization. The presence of the group's distinctive pollen further hints that the Cheirolepidiaceae may have survived into the Paleocene in some parts of the Northern Hemisphere.

MORPHOLOGICAL FEATURES The Cheirolepidiaceae included trees up to 3 m (10 ft) in diameter. All members of this group were united in producing a unique type of pollen grain called *Classopollis*. *Classopollis* pollen was spherical, with bands around the equator where the pollen wall thinned. Some *Classopollis* species also had a small lid at their pole. The pollen type is so distinctive that it makes this group, with an otherwise slim fossil record, easy to track through time.

Pseudofrenelopsis parceramosa, *Protopodocarpoxylon* wood, and *Classostrobus comptonensis*

LOCATION Wealden, Isle of Wight, England.
AGE Early Cretaceous, late Barremian to early Aptian (129–120 million years ago).
CHARACTERISTICS Unbranched shoots a few centimeters long found in thick mats. Nodes developed every 0.7–1 cm (0.3–0.4 in) and leaves completely encased the shoot, with only a small pointed tip free. Leaves had distinct parallel ridges, thick outer coverings, and small hairs, all features associated with conserving water. *Protopodocarpoxylon*–type wood with growth rings found with the foliage. *Classostrobus comptonensis*–type pollen cones were spherical and ranged from 1–1.7 cm (0.4–0.7 in) in diameter. Pollen sacs hung from a tiny umbrella-shaped structure in which the handle of the umbrella attached to the core of the cone. *Classopollis*–type pollen found in surrounding sediment. Seed-bearing cones have not been described from this locality, but in similarly aged rocks from Korea, elliptical seed cones 4.3 cm (1.7 in) long and 2 cm (0.8 in) wide have been found with *Pseudofrenelopsis parceramosa* foliage.

FOUND IN Gray silty clay of the Brook Formation. Fossil plants concentrated in several organic-rich layers that may have represented forest litter.
HABITAT Streamside forest.
NOTES *Pseudofrenelopsis parceramosa* has also been found in the United States, Portugal, central Europe,

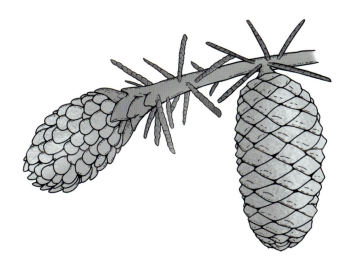

Pseudofrenelopsis parceramosa seed cone (right) and *Classostrobus comptonensis* pollen cone (left).

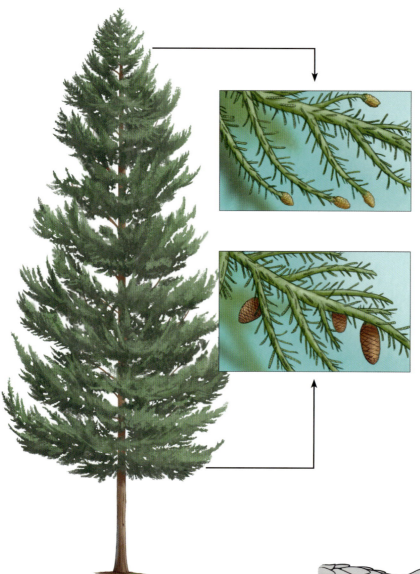

The tree bearing *Classostrobus comptonensis* pollen cones (top) and *Pseudofrenelopsis parceramosa* seed cones (bottom).

Korea, and Africa in rocks ranging in age from Berriasian to early Cenomanian, making it one of the most widespread foliage types produced by this group. Some of the twigs found at Wealden showed a pattern of spiral cracking typical of soft rot, and some contained preserved fungal hyphae. This suggests that the twigs lay on the forest floor for some time before being buried and preserved.

Tomaxellia biforme

LOCATION Santa Cruz Province, Patagonia, southern Argentina.
AGE Early Cretaceous, Aptian (125–113 million years ago).
CHARACTERISTICS Two sizes of leaves found together surrounding the tips of branches. Small leaves were up to 5 mm (0.2 in) long and 3 mm (0.1 in) wide, scale-shaped, pressed tightly to the stem, and had a pointed free tip. Larger leaves were up to 2 cm (0.8 in) long and 3 mm (0.1 in) wide, with the lower half of the leaf pressed to the branch and the upper half curving away. Leaves had a semicircular or triangular cross section with a keel-like structure facing outward. The transition between small and large leaves was gradual. Oblong seed cones attached to the tips of foliage branches were up to 3 cm (1.2 in) long and 1.5 cm (0.6 in) wide. Seeds developed on structures in which the bract and scale were fused.

Tomaxellia biforme with immature (left) and mature (right) seed cones.

Bract-scales were arranged in a compact spiral around a spongy core. Tiny pollen cones found attached alone on the sides of branches or in pairs at the tip. Pollen cones were about 3 mm (0.1 in) long and 1 mm (0.04 in) wide and consisted of small, circular sacs containing *Classopollis* pollen.

FOUND IN Mudstones of the Anfiteatro de Ticó Formation.

HABITAT Floodplain and lakeside forests that included streamside horsetails, an understory of ferns and cycads, and a canopy that included several species of conifers and ginkgos.

NOTES At first glance, the foliage of *Tomaxellia biforme* resembles several contemporaneous conifers, including *Sequoia* and *Taxodium*. However, details of the stomata reveal that it is distinct.

Classopollis pollen grains were about 30 μm in diameter.

CUPRESSACEAE

The cypresses first appeared in the Middle Jurassic and diversified modestly through the Cretaceous. They continued producing species through the Cenozoic to become the most diverse living conifer group, with about 30 genera and more than 130 species.

MORPHOLOGICAL FEATURES Mesozoic members of the Cupressaceae were trees. Most were evergreen but some, including species of *Glyptostrobus*, *Metasequoia*, and *Taxodium*, shed their foliage seasonally. Bark on mature trees is commonly reddish brown and stringy or flaky. Scale-like leaves are arranged in whorls, a spiral, or in pairs that alternate at 90° angles.

Elatides thomasii

LOCATION Yorkshire, England.

AGE Middle Jurassic, Bajocian to Callovian (170–164 million years ago).

CHARACTERISTICS Leaves up to 5 mm (0.2 in) long and 1 mm (0.04 in) wide, arranged in a spiral around slender branches. Leaves square in cross section, sickle-shaped, and tapered to a sharp point. Pollen cones 1.5–1.7 cm (0.6–0.7 in) long and 3–4 mm (0.1–0.2 in) wide. Pollen sacs attached to umbrella-shaped bracts with a rhomboidal head about 1 mm (0.04 in) wide. Spherical seed cones about 2 cm (0.8 in) in diameter attached to the tips of leafy shoots. Seeds developed under umbrella-shaped bracts that were up to 4 mm (0.2 in) wide and 2 mm (0.08 in) high.

FOUND IN Dark-gray clay of the Hasty Bank locality, Saltwick Formation.

HABITAT Floodplain forest.

NOTES Size distinguishes *Elatides thomasii* from *E. williamsonii*, which occurs in the Gristhorpe Plant Bed. All parts of *Elatides thomasii* were smaller than those of *E. williamsonii*. Separating species based on size can be risky. Plants vary the size of their organs based on age, climate, and growing conditions. However, cone size tends to change less with environmental conditions, and the size ranges for *Elatides thomasii* and *E. williamsonii* cones do not overlap, adding confidence to the identification of separate species.

Elatides zhoui

LOCATION Tugrug, Mongolia.

AGE Early Cretaceous, Aptian to Albian (125–100 million years ago).

CHARACTERISTICS Leafy shoots branched alternately. Leaves arranged in a spiral around branches. Individual leaves about 5 mm (0.2 in) long, oval, and tapering to a slightly hooked tip. Pollen cones developed on the sides of branches in clusters that spiraled around the branch. Each cone had a slender axis to which small triangular scales attached by a small stalk. Pollen sacs developed at the base of each scale. Seed cones formed at the tip or sides of leafy branches. Seed cones oblong, about 1.5 cm (0.6 in) in length, with a woody axis and triangular bracts attached in a spiral. Edges of each bract were toothed and held four to six seeds. Seeds slightly flattened, with a narrow wing.

FOUND IN Lignite seams in an open pit mine within the Khukhteeg Formation, which also includes conglomerates, sandstone, and siltstone associated with a fast-flowing river.

HABITAT Floodplain forest.

Pollen cone (left) and seed cone (right) of *Elatides zhoui*.

NOTES Evolutionary analysis shows that *Elatides*, including *E. zhoui*, *E. williamsonii*, and *E. thomasii*, were closely related to living *Cunninghamia*, which first appeared in the Middle Jurassic and survives today in China, Taiwan, Vietnam, and Laos.

Metasequoia occidentalis

LOCATION Found throughout the middle and high latitudes of the Northern Hemisphere.

AGE Late Cretaceous, Cenomanian (100 million years ago) to present (as *Metasequoia glyptostroboides*).

CHARACTERISTICS *Metasequoia* makes a wide, buttressed trunk with stringy, reddish bark. Leaves on branchlets that are arranged opposite to each other and shed as a unit at the beginning of the cold season. Leaves 1–3 cm (0.4–1.2 in) long and arranged opposite one another on branchlets. Seed cones spherical and about 2 cm (0.8 in) in diameter. Sixteen to 28 bracts arranged in opposite pairs. Five to eight seeds developed on each scale. Pollen cones about 6 mm (0.2 in) long and produced in pairs between foliage leaves. Pollen globular with a distinct pointed nub on the apex.

Elatides zhoui.

Metasequoia occidentalis with seed cones.

FOUND IN In the Cretaceous, most *Metasequoia occidentalis* fossils were preserved in floodplain mudstones and sandstones, as well as in lake sediments.

HABITAT Mesozoic *Metasequoia* grew across a wide range of habitats, including floodplains and upland forests to wetland margins.

NOTES About 20 species of fossil *Metasequoia* have been described from the Late Cretaceous and Cenozoic of the Northern Hemisphere. In a comprehensive review of these, Yan-Ju Liu concluded that all but the middle Eocene species *Metasequoia milleri* (radioisotopically dated to 48.7 million years ago) and *M. foxii* (late Paleocene, 59–56 million years ago) could be circumscribed by the variation observed in living species *M. glyptostroboides*. Liu recommended that all these fossils be drawn under the umbrella name *Metasequoia occidentalis*. Liu further concluded that fossil *Metasequoia occidentalis* and living *M. glyptostroboides* were probably the same long-lived species, but to avoid confusion, she advocated keeping the names of fossil and living members separate.

Metasequoia pollen grains were about 30 μm in diameter.

PODOCARPACEAE

Today the Podocarpaceae grow mostly in the Southern Hemisphere, but their Mesozoic roots were global. They diverged from the Araucariaceae in the Permian. Recognizing the fossil representatives has been difficult because living members have varied form and have changed significantly since the Mesozoic.

MORPHOLOGICAL FEATURES Tall trees and bushy shrubs. Many, but not all, podocarps produce pollen and seed cones on separate individuals. Leaves are generally arranged in a spiral and may be elongated, elliptical, or scale-like. Pollen cones dangle from the sides or tips of branches and produce numerous pollen sacs. Pollen has an air-filled bladder for buoyancy. Seed cones vary tremendously among the living podocarps, although most are fleshy and displayed on a stalk.

Mataia podocarpoides

LOCATION The Clent Hills, Canterbury Region, South Island, New Zealand.

AGE Late Jurassic, Tithonian (152–145 million years ago).

CHARACTERISTICS Twigs with leaf-covered branches. Individual leaves arranged in a spiral, with each leaf twisted to be displayed in a flat surface. Leaves 5–15 mm (0.2–0.6 in) long and 2–4 mm (0.08–0.2 in) wide, with a single prominent vein and pointed tip. Seed cone on a short stalk was more than 2 cm (0.8 in) long. A fleshy scale surrounded the seed, which was about 1.5 mm (0.06 in) long and 1 mm (0.04 in) in diameter.

FOUND IN Sandstones deposited by a river associated with the Clent Hills Group.

HABITAT Floodplain or river margin forest.

NOTES *Mataia podocarpoides* has also been found in the Malvern Hills locality in Canterbury, New Zealand, and similar-age rock near Walloon in Queensland, Australia, demonstrating that it was widely distributed across the region during the Late Jurassic.

Mataia podocarpoides foliage.

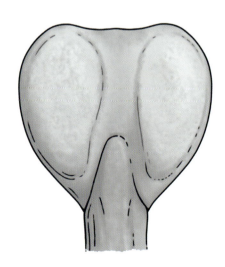

Mataia podocarpoides seed-bearing structures.

Mataia podocarpoides.

Off the west coast of South Africa, entire Cretaceous forests
are preserved on the seafloor.

Podocarpoxylon jago

LOCATION Fifty km (32 mi) offshore of the Buffels River, Namaqualand, South Africa.

AGE Late Cretaceous, Coniacian (88–83 million years ago).

CHARACTERISTICS Fossil trunks were found in so-called "life position"—the position and orientation in which they originally grew. Some stumps stood upright, while others lay where they fell tens of millions of years ago. The ancient podocarp tree that yielded the fossil specimen was at least 50 cm (1.6 ft) in diameter. Wood lacked growth rings, suggesting that the tree grew in a climate without seasons. This species differed from other podocarp wood in having a distinctive series of thin-walled cells scattered among the tracheids. These so-called ray cells allowed the plant to move nutrients or waste from the living part of the wood into the interior of the trunk.

FOUND IN Sediments originally deposited at the mouth of an ancient river during a time of lower sea level.

HABITAT Coastal forest.

NOTES When paleobotanists name fossil plants, they choose names that are somehow meaningful or memorable. In this case, the genus name was predetermined. The fossil was definitely "podocarp-like wood," or *Podocarpoxylon*. However, Marion Bamford and Ian Stevenson had choices about the second part of the name. They might have chosen a word that described an important feature of the fossil or invoked the place where it was found. They might have chosen to honor a person. Bamford and Stevenson named their new fossil *Podocarpoxylon jago* after the *Jago* mini-submersible that made it possible to collect fossils from 140 m (460 ft) below the ocean surface.

Two other species of podocarp wood were recovered from the submerged forest: *Podocarpoxylon umzambense* and another, as yet unnamed, species. *Podocarpoxylon umzambense* was previously discovered in another submerged forest about 100 km (62 mi) to the north and on land along South Africa's east coast. These finds show that a diverse podocarp forest occupied the region during the Late Cretaceous.

Rissikia media

LOCATION Burnera Waterfall, eMkhomazi, KwaZulu-Natal, South Africa.

AGE Late Triassic, Carnian (237–227 million years ago).

The wood of *Podocarpoxylon jago*.

Rissikia media with seed cones.

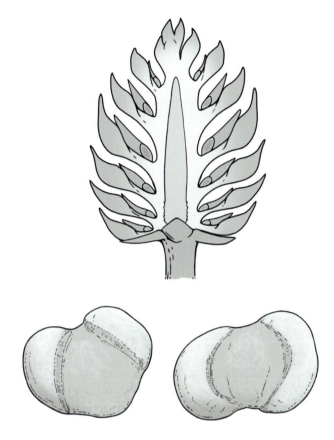

ABOVE LEFT: *Rissikia media* seed cone in cross section, revealing seeds.

ABOVE RIGHT: *Rissikia media* pollen cone.

LEFT: Pollen associated with this conifer was about 90 μm across.

CHARACTERISTICS Foliage consisted of 6 cm (2.4 in) long branchlets. Branchlets had small, scalelike leaves at the base and larger leaves that emerged in a spiral and twisted to form a row on either side of the branchlet, making a flat, light-gathering surface. Individual leaves about 7 mm (0.3 in) wide and of variable length depending on where along the branchlet they emerged. Leaves thick and rhomboidal in cross section and with a pointed tip. Larger branches covered in spiky leaves pressed close to the branch. Pollen cone elliptical, about 1 cm (0.4 in) long and 5 mm (0.2 in) wide, and attached to the branch by a sturdy stalk. Pollen cones consisted of about 25 umbrella-shaped bracts, each of which bore two pollen sacs. Pollen had two air-filled bladders for buoyancy. Seed cone produced on the tips of leafy branches. Seed cone about 3 cm (1.2 in) long and spikelike, with foliage leaves grading into non-photosynthetic, lobed bract-scales that produced seeds on a short stalk.

FOUND IN Mudstones of the Molteno Formation.

HABITAT Floodplain forest.

NOTES John Townrow proposed that *Rissikia* might explain the evolution of the fleshy scale that surrounds the podocarp seed, speculating that it evolved from one of the lobes of the *Rissikia* bract-scale that rolled around the seed and became fleshy.

TAXACEAE

Yews are evolutionary sisters of the Cupressaceae that first appeared in the Early Jurassic. The group diversified in the Jurassic and retreated to marginal environments during the Flowering Plant Revolution. Today, species of the Taxaceae are found in temperate regions of the Northern Hemisphere.

MORPHOLOGICAL FEATURES Evergreen conifers that grow as small, branchy trees or shrubs. Sturdy, flattened leaves with a strong central vein develop on a short stalk. Leaves emerge from branches in a spiral, then twist to arrange themselves on either side of the branch to display a flat surface to the sun. Leaves vary in size depending on species and whether the plant grows in sun or shade. Flimsy pollen cones are only a few millimeters long and produce pollen sacs on paperlike scales. Members of the Taxaceae produce a single scale that displays one or two ovules. Maturing seeds are covered in a brightly colored, fleshy structure that entices birds to disperse the hard seeds within. If these fleshy "yew berries" were present in the earliest Jurassic forms, they preceded birds onto the evolutionary stage and may have evolved to attract other dispersers. Yew seeds are toxic to mammals!

Marskea heeriana

LOCATION Irkutsk Coal Basin, East Siberia, Russia.
AGE Middle Jurassic, Aalenian to Bajocian (174–168 million years ago).

Marskea heeriana. Only leaves are known.

CHARACTERISTICS Leaves 4–10 mm (0.2–0.4 in) long and 1–3 mm (0.04–0.1 in) wide, attached to branches by a short stalk and arranged in a spiral around branches. Leaves twisted to display a single plane to the sun. No reproductive structures have been found.
FOUND IN Organic-rich siltstones of the Prisayan Formation.
HABITAT Streamside forest.
NOTES Although neither seeds nor pollen cones have yet been found, distinctive features of the leaf surface, including the shape of cells and the presence of minute hairs surrounding the stomata, unite this species with the Taxaceae.

Palaeotaxus rediviva

LOCATION Southern Sweden.
AGE Early Jurassic, Hettangian (201–199 million years ago).
CHARACTERISTICS Straight leaves, 1.2–2.2 cm (0.5–0.9 in) long and about 2 mm (0.08 in) wide, with a strong central vein, arranged in a spiral along the branch. Leaves twisted at their bases to lay in a single plane. Seed-producing shoots arose along branches and were covered in spirally arranged scales lying flat against the branch. The fertile shoot ended in a single

ovule that pointed away from the plant. Pollen and pollen cones are unknown.

FOUND IN Clays of the Bjuv Member of the Höganäs Formation.

HABITAT Swamp forest.

NOTES The leaves of *Palaeotaxus rediviva* clearly showed that it was a member of the Taxaceae. However, the seeds were squashed, and Rudolf Florin could not tell whether the species possessed the distinctive fleshy seed covering characteristic of the Taxaceae. The timing of the evolution of this key feature remains a mystery.

Taxus guyangensis

LOCATION Guyang County, Inner Mongolia, China.

AGE Early Cretaceous, Aptian to Albian (125–100 million years ago).

CHARACTERISTICS Leaves covered most of the branch. Individual leaves were 1.8–3.2 cm (0.7–1.3 in) long and 2–3 mm (0.08–0.1 in) wide, arranged in a spiral on short stalks that twisted to display leaves in a single plane. Leaves had an obvious central vein and a pointed tip. Seeds developed in pairs on the tip of leafy branches. Three pairs of leaflike scales arranged at right angles to the pair below occurred at the base of each ovule. In some specimens, one ovule was mature, with a surrounding fleshy covering, and the other was not. This may indicate that only one was pollinated or that ovules were developmentally timed to mature asynchronously. Seeds were ovoid,

Taxus guyangensis.

7 mm (0.3 in) long, and 6 mm (0.2 in) wide.

FOUND IN Organic-rich oil shales of the Guyang Formation.

HABITAT Coastal forest.

NOTES Two other species of *Taxus* have been described from the Early Cretaceous, *Taxus intermedius* and *T. acuta*. The relatively low diversity of this group in the fossil record may indicate that many of its members lived in places where they were unlikely to be preserved. Alternatively, the genus's modern diversity may have evolved after the Mesozoic.

VOLTZIALES

This extinct group originated during the Pennsylvanian and gave rise to all the Mesozoic conifers. It includes two major evolutionary branches: the walchian type, named for *Walchia*, known from the Late Pennsylvanian to Early Permian (310–290 million years ago), and the voltzian type, named for *Voltzia*, known from the Permian and Triassic. The voltzian conifers displayed features intermediate between Pennsylvanian and Permian forms and the modern groups of the Mesozoic. However, evolutionary analysis shows that conifer evolution was not linear and progressive. Instead, many species experimented with different combinations of features, with the variety of Mesozoic conifers representing those lineages that happened to survive. The Voltziales became extinct during the Cretaceous angiosperm revolution.

MORPHOLOGICAL FEATURES Voltziales were trees with a straight trunk and branches emerging in whorls. Voltziales produced a variety of leaf forms, ranging from thick, hook-shaped leaves that clung to the branch to narrow, strap-like leaves with robust parallel veins. Leaves were arranged in a spiral around branches. Branches tended to grow upward with leaves flattened and arranged on small side branches. Seeds of voltzian conifers developed on many-lobed scales clustered into a specialized cone, an innovation that linked the Voltziales to conifers that emerged in the Triassic and Jurassic.

Krassilovia mongolica and *Podozamites harrisii*

LOCATION Tevshiin Govi, central Mongolia.
AGE Early Cretaceous, Aptian to Albian (125–100 million years ago).
CHARACTERISTICS Leaves oblong to strap-shaped, ranging from 1–5 cm (0.04–2.0 in) long and 5–10 mm (0.2–0.4 in) wide, and arranged in a spiral on shoots that may have been shed regularly. Leaves had rounded tips, tapered bases, and attached to branches by short stalks that twisted to display foliage in a single plane. Leaves had many closely spaced parallel veins. Seed cones spherical to oblong, 15–20 cm

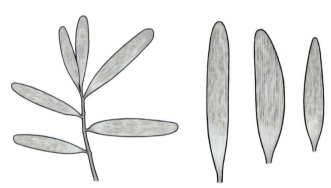

Krassilovia mongolica foliage.

Krassilovia mongolica.

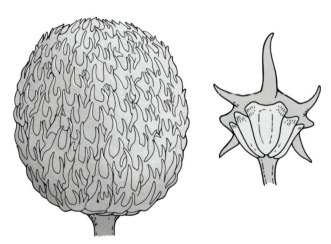

Krassilovia mongolica seed cone (left) and bract with seeds (right).

(5.9–7.9 in) long and about 15 cm (5.9 in) wide. Cones composed of overlapping, umbrella-shaped scales that attached to the central core of the cone by short stalks. Cone scales interlocked with five or six long spikes that emerged from their edges. Up to five ellipsoid seeds developed on each scale. Seeds were about 2 mm (0.08 in) long, with a pair of thin wings to aid dispersal.

FOUND IN Brown coals of the Tevshiin Govi Formation.

HABITAT Swamp forest.

NOTES This is among the youngest member of the Voltziales yet reported and is united with the group by details of the stomata and surface cells on the leaf.

Patokaea silesiaca with Brachyphyllum and Pagiophyllum foliage

LOCATION Patoka Clay Pit, near Lubliniec, Poland.

AGE Late Triassic, Norian (227–208 million years ago).

CHARACTERISTICS Leaf-covered shoots that branched in a single plane. Branches covered with spirally arranged rhomboidal leaves 4–6 mm (0.2 in) long. The leaf pressed tightly to the branch or curved outward. The tip of the leaf rounded or pointed. Seed cones fell apart at maturity, and most of the fossils are of single cone scales. Scales ranged from 5–13 mm (0.2–0.5 in) long, with a stalk. Scales divided into three tongue-shaped lobes, the outer two of which slightly curved inward to partly surround a single ovule on each lobe. Ovules about 4.5 mm (0.2 in) long and 2 mm (0.08 in)

Patokaea silesiaca seed.

Patokaea silesiaca.

155

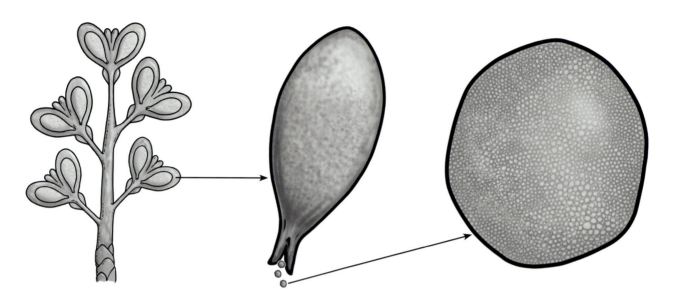

Patokaea silesiaca seed cone, ovule, and pollen.

wide, with a 1 mm (0.04 in) long neck and forked tips for capturing pollen. Round, rough-textured pollen grains were preserved within the neck of some ovules. Fully developed seeds were up to 8.5 mm (0.3 in) long and 7 mm (0.3 in) wide, slightly egg-shaped, and flattened. Pollen cones developed at the end of leafy twigs. Cones oval in outline, up to 6 mm (0.25 in) long and 4 mm (0.2 in) wide. Bracts containing clusters of pollen sacs were arranged in a spiral.

FOUND IN Mudstones and siltstones of the Patoka Member of the Grabowa Formation.

HABITAT Floodplain or swamp forest.

NOTES The leaf forms *Brachyphyllum* and *Pagiophyllum* are distinguished from one another by the proportion of the leaf that is free from the branch. In *Brachyphyllum*, the free part of the leaf is about the same length as the portion of the leaf attached to the branch. In *Pagiophyllum*, the free part of the leaf is longer than the attached part. The Patoka Clay Pit fossils show that both types of foliage can occur on the same plant and even on the same branch. *Brachyphyllum* and *Pagiophyllum* were common Mesozoic leaf forms that have been found attached to branches of many groups, including Araucariaceae and Cheirolepidiaceae.

Voltziopsis africana

LOCATION New South Wales, Australia.
AGE Probably Early Triassic but possibly latest Permian (260–237 million years ago).

Voltziopsis africana foliage.

CHARACTERISTICS Leaves of various shapes and sizes appeared on the same branching shoot. The smallest leaves were scalelike, about 3 mm (0.1 in) long and 2 mm (0.08 in) wide, approximately triangular, with a pointed end. Large leaves were up to 3 cm (1.2 in) long and 5 mm (0.2 in) wide, displayed on the sides of the branch. Large leaves tended to twist so that they created a flattened surface to capture light. Seed cones developed on the tips of short side branches. Cones were elliptical, about 2.5 cm (1 in) long and 1.5 cm (0.6 in) wide, and composed of about 25 scales packed closely together. Each scale widened away from the branch and divided into five or six lobes, with one seed developing on each. Pollen and pollen cones have not been discovered.

FOUND IN Shales above coal of the Bulli Seam, part of the Illawarra Coal Measures in the Sydney Basin.

HABITAT Swamps and wetlands.

NOTES John Townrow described *Voltziopsis africana* from fossils that he did not collect himself. In his 1967 paper he writes, "In New South Wales the foliage ascribed to *V. africana* comes only, as far as I can discover, from the roof shales of the Bulli Seam." This coal lies very near the boundary between the Permian and Triassic periods, the geologic moment of the end-Permian extinction event. High-precision age dating of rocks in the Sydney Basin shows that the transition from the Permian to the Mesozoic was recorded in clays just above the Bulli Seam. Therefore, if Townrow's guess on the location of the fossil is correct, it could be among the earliest Triassic conifers. *Voltziopsis africana* was originally described from Early Triassic sediments in present-day Tanzania, indicating that the species spread across Gondwana during the Early Triassic.

ENIGMATIC SEED PLANTS WITH MESOZOIC ROOTS

These plants occupied various branches of the evolutionary tree and—apart from the Gnetales—did not survive the Cretaceous angiosperm revolution. Several possessed features that contributed to angiosperm success, but none of these plants *were* angiosperms. Instead, they testify to the evolutionary pressures—disturbance, climate change, the explosion of animal pollinators and dispersers—that molded the angiosperms into the ecological juggernaut that they became.

BENNETTITALES

The Bennettitales arose in the Early Permian, survived the end-Permian extinction event, and became among the most common plants of the Mesozoic. Like many others, they began a slow slide to extinction in the middle of the Cretaceous and were probably gone before the asteroid punctuated the end of the Mesozoic.

MORPHOLOGICAL FEATURES The Bennettitales grew as slender shrubs, stout branching forms, and ground-hugging botanical beach balls. They shared foliage types with cycads, and except for some details of the stomata, Bennettitalian and cycad leaves were difficult to distinguish. Some Bennettitales produced separate seed and pollen cones. Others produced flowerlike reproductive structures in which ovules were surrounded by pollen-producing parts that were, in turn, displayed in a whorl of showy bracts.

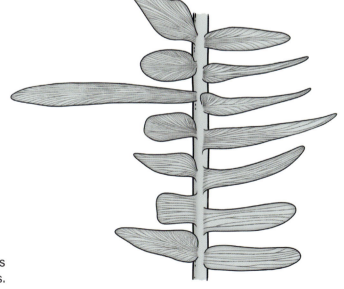

OPPOSITE: **Landscape of the Triassic (Norian) of Arizona, USA.**

RIGHT: **The diversity of pinnule types found in the Bennettitales.**

Cycadeoidea megalophylla

LOCATION Isle of Portland, England.

AGE Late Jurassic, Tithonian (152–145 million years ago).

CHARACTERISTICS Stems were irregularly shaped, flattened spheres of various sizes. The original specimen was about 30 cm (11.8 in) in diameter and 15 cm (5.9 in) high. The surface was covered with rhombohedral leaf bases about 5 cm (2 in) wide and 2 cm (0.8 in) long and represented locations where leaves attached. Some bases contained the remains of leaves, while others were preserved as open spaces surrounded by sturdy scales. Reproductive structures scattered among the leaf bases looked like knots on the surface. In cross section these included a central fleshy core upon which ovules developed, surrounded by fingerlike projections that produced pollen sacs. The entire structure was encased in scales much like those that surrounded the leaf bases.

FOUND IN Fossil soil of the Lulworth Formation (formerly the Lower Purbeck Beds).

HABITAT *Cycadeoidea* trunks were found in life position in a fossil soil layer. The fossil-bearing layer occurs just above a lagoonal limestone, suggesting a coastal habitat.

NOTES Reproductive structures of *Cycadeoidea megalophylla* produced pollen and seeds in the same structure, a feature that some use to connect this group to flowering plants. More intriguing is how these

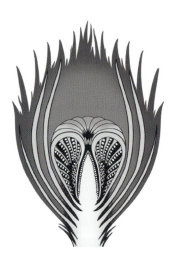

plants were pollinated. All appeared tightly closed, leading some to suggest that they self-pollinated. However, tiny drill holes and insect waste left behind by feeding insects indicated that insects entered the reproductive structures, perhaps as predators as well as pollinators.

Reproductive structure of *Cycadeoidea megalophylla*.

Wielandiella angustifolia

LOCATION Scania, southern Sweden.

AGE Late Triassic, Rhaetian, to Early Jurassic, Hettangian (208–199 million years ago).

CHARACTERISTICS Woody, shrub-sized plant with a spreading canopy. Stem covered in rhomboidal leaf scars from which leaves dropped during life. Leaves, *Anomozamites angustifolius*, up to 25 cm (9.8 in) long and 3 cm (1.2 in) wide, with a robust central axis and leaflets that were about 1.4 cm (0.6 in) long and 8 mm (0.3 in) wide. A flange of photosynthetic tissue developed on either side of the petiole. In addition to these large leaves, smaller versions occurred in the same beds, suggesting variation in leaf size on different parts of the plant. Smaller, entire margin leaves encircled the seed-producing structures. Similar leaves, termed bracts because they lacked the strong central rib, surrounded the oval seed-producing structure. Seed heads developed in the space between branches. Each was 4.4 cm

Cycadeoidea megalophylla.

(1.7 in) long and 1.5 cm (0.6 in) wide and covered in ovules that were each surrounded and protected by sturdy scales.

FOUND IN Höör Sandstone and Bjuv Member of the Höganäs Formation.

HABITAT Open habitat along rivers or the coast.

NOTES The showy bracts around the seed-producing structure suggest an insect pollination partner. However, only a few of the seeds matured in each head, which may mean that pollination was ineffective.

Williamsonia sewardiana

LOCATION Rajmahal Hills, Bihar, India.

AGE Late Jurassic, Bajocian to Oxfordian (170–167 million years ago).

CHARACTERISTICS Stout stem up to 2 m (6.5 ft) tall and 10 cm (3.9 in) in diameter, with smaller branches emerging at irregular intervals. The plant produced a small amount of wood but most of its stem strength came from an armor of sturdy leaf bases that remained attached to the outside of the stem

Wielandiella angustifolia.

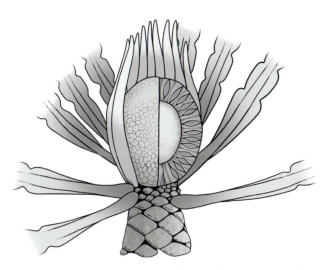

The reproductive structure of *Wielandiella angustifolia* with interior structure revealed.

Williamsonia sewardiana.

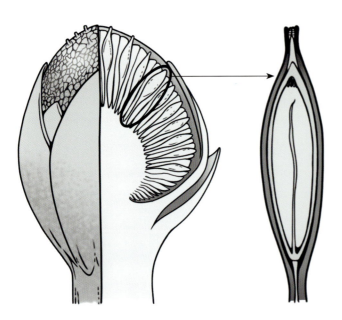

Williamsonia sewardiana seed-bearing structure (left) and ovule detail (right).

Williamsonia sewardiana pollen-producing structure.

after the leaves had perished. The growing tip of the plant was surrounded by woody bracts that protected developing leaves. Leaves of this species were called *Ptilophyllum cutchense*, a once-pinnate type with a thick central stem and broad, rounded pinnules. The seed-producing structure developed on the ends of

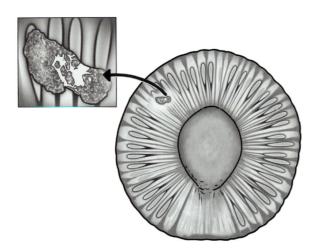

Williamsonia bockii. Damage like this to reproductive structures in some Bennettitales suggests that insects entered and may have pollinated this group.

lateral branches. Seeds produced on a hemispherical central core covered in ovules on stalks; each ovule was surrounded by a series of flat-topped scales that fit together to form a protective shield so that only the ovule entrance was exposed. The seed-producing structure was surrounded by several whorls of hairy bracts that formed a conical shape around the ovule-bearing structure.

FOUND IN Cherts likely formed when mineral-rich water penetrated peat deposits.

HABITAT Wetlands or lake margins.

NOTES Fossil reproductive structures commonly contain seeds at various stages of development. This may indicate poor pollination. It could also mean that these structures were available to pollinators for extended periods and seeds remained on the plant awaiting dispersal.

Williamsoniella coronata with *Nilssoniopteris vittata* foliage

LOCATION Gristhorpe Plant Bed, Yorkshire, England.
AGE Middle Jurassic, Bajocian (170–168 million years ago).

Williamsoniella coronata.

CHARACTERISTICS Slender, branching stems with smooth bark and leaves irregularly scattered along the shoot. *Nilssoniopteris vittata* leaves were up to 20 cm (7.9 in) long and 7 mm (0.3 in) wide and shaped like an elongated oval. Margin was entire with a strong midrib from which smaller veins emerged at right angles and commonly branched once before they reached the leaf margin. The leaf tip was rounded. Reproductive structures included both ovules and pollen organs and developed in the angle between branches. Reproductive structure developed on a 3 cm (1.2 in) long stalk that swelled into a pear shape, with the narrow end topped by a crown-like ornament. The swollen part of the stalk was covered in ovules surrounded by scales that broadened at their top to cover all the ovule except the opening where pollen entered. Below and around this structure, a whorl of wedge-shaped pollen organs each produced five or six pollen sacs. Five petal-like leaves, 1.5 cm (0.6 in) long and 3 mm (0.1 in) wide, encircled the reproductive structure.

FOUND IN Clays of the Cloughton Formation.
HABITAT Floodplain forest.
NOTES The flowerlike reproductive structures of *Williamsoniella coronata* focused attention on this group as a possible angiosperm relative. *Williamsoniella* had flowers that, like angiosperms, produced both ovules and pollen. In 1907, this observation led Edward Newell Arber and John Parkin to propose that plants including *Williamsoniella*, along with *Pentoxylon*, *Caytonia*, and living *Gnetum,* formed an evolutionary group with flowering plants: they called this new group the "anthophytes." They argued that the anthophytes connected Mesozoic seed plants and mid-Cretaceous flowers. However, by the end of the of the twentieth century, genetic evidence removed *Gnetum* to sisterhood with conifers, and close study of Mesozoic fossils eliminated each from contention for angiosperm ancestor. It appeared that the many flowerlike reproductive structures evolved independently from seed plant stock, probably in response to an explosion of new pollinators.

CAYTONIALES

The Caytoniales populated wetland ecosystems worldwide from the Middle Triassic through the Early Cretaceous.

MORPHOLOGICAL FEATURES The Caytoniales were shrubs or perhaps small trees that may have shed leaves seasonally. Leaves, called *Sagenopteris*, were compound, with four leaflets attached like the fingers of a hand to a thick petiole. Leaflets had netlike veins resembling those of flowering plants. The ovules of the Caytoniales were surrounded by a fleshy covering that mimicked angiosperm fruit, but unlike angiosperms, the seeds were not completely enclosed in this second layer. Pollen sacs dangled in unprotected clusters on specialized branches, suggesting that the plant relied on wind pollination.

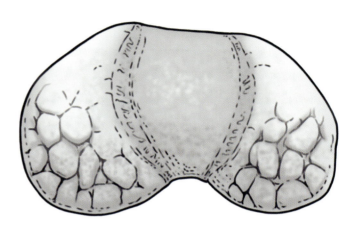

Caytonia pollen grain.

Caytonia nathorstii, Caytonanthus oncodes, and *Sagenopteris colpodes*

LOCATION Gristhorpe Plant Bed, Yorkshire, England.
AGE Middle Jurassic, Bajocian (170–168 million years ago).
CHARACTERISTICS Compound leaves in which four leaflets were arranged like the fingers of a hand on a stout petiole. Leaflets were about 2.5 cm (1 in) long and 1.5 cm (0.6 in) wide, with a strong central rib from which smaller veins divided and rejoined in a network. Leaflet tips were pointed. *Caytonia nathorstii* seed-bearing structures consisted of a ribbed stalk about 3 cm (1.2 in) long, with six to eight pairs of seed-producing organs arranged on opposite sides of the stalk. Each included eight ovules enwrapped in a thick protective flap that became fleshy when the seeds matured. Pollen organs, *Caytonanthus oncodes,* consisted of

short stalks 2–3 cm (0.8–1.2 in) long that held a cluster of four-part pollen sacs each 2–3 mm (0.08–0.1 in) long. *Caytonanthus oncodes* produced pollen with two air-filled bladders, much like that of modern pines.

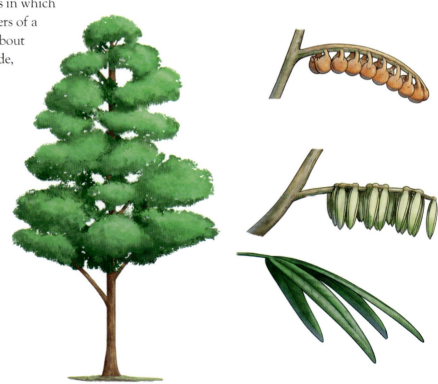

Caytonia nathorstii with seed-bearing structures (top right), pollen organs (middle), and foliage (bottom).

FOUND IN Floodplain shales of the Cloughton Formation.

HABITAT Floodplain forest with poorly drained soils.

NOTES Closely related species may share leaf features even when other parts have evolved differences. For example, foliage of *Sagenopteris colpodes* in Yorkshire came in two sizes. The larger form was always associated with *Caytonia kendallii* seeds, while the smaller form was always found with *Caytonia nathorstii*. Other than the size difference, *Sagenopteris colpodes* leaves were indistinguishable.

Caytonia sewardii, Caytonanthus arberi, and Sagenopteris phillipsii

LOCATION Cayton Bay, Yorkshire, England.

AGE Middle Jurassic, Bajocian (170–168 million years ago).

CHARACTERISTICS Compound leaves in which four leaflets were arranged like the fingers of a hand on a stout petiole. Leaflets were 3–4 cm (1.2–1.6 in) long and 0.5 cm (0.2 in) wide, with a strong central rib and net-type veins. Leaflet tips were pointed. *Caytonia sewardii* seed-bearing structures consisted of a smooth stalk about 3 cm (1.2 in) long, with six to eight pairs of seed-producing organs arranged on opposite sides of the stalk. Each seed-producing

organ contained six to nine ovules surrounded by a flap of tissue that became fleshy when the seeds matured. Pollen organs, *Caytonanthus arberi*, consisted of short stalks 2–3 cm (0.8–1.2 in) long. Each stalk held four-part cylindrical pollen sacs, each 3–4 mm (0.1–0.2 in) long, which produced pollen with two air-filled bladders.

FOUND IN Organic-rich shales of the Cloughton Formation.

HABITAT *Sagenopteris* foliage commonly occurred with familiar wetland plants like horsetails, suggesting a preference for marshy habitats.

NOTES *Sagenopteris*'s network of veins coupled with the double-layered seeds of *Caytonia* placed a spotlight on this species as a possible angiosperm ancestor. Moreover, the Middle Jurassic age was about right for the emergence of flowering plant features. However, *Caytonanthus* and its very conifer-like pollen soon removed the group from contention. *Caytonia* was just another Jurassic seed plant experimenting with flowerlike features.

Sagenopteris trapialensis with Caytonia and Caytonanthus

LOCATION Cañadón Asfalto Basin, Chubut Province, Patagonia, Argentina.

LEFT: *Caytonia sewardii* seed-bearing structure. MIDDLE: *Caytonanthus arberi* pollen organ.
RIGHT: *Sagenopteris phillipsii* foliage.

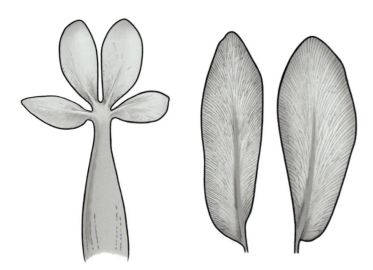
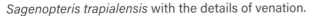

Sagenopteris trapialensis with the details of venation.

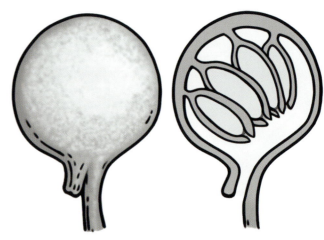

Caytonia seed-bearing structure from Patagonia including cross section showing four ovules.

AGE Early Jurassic, Pliensbachian to early Toarcian (190–178 million years ago).

CHARACTERISTICS *Sagenopteris trapialensis* leaves consisted of four leaflets arranged like fingers on a hand on a 3 cm (1.2 in) long petiole. Individual leaflets were up to 6 cm (2.4 in) long and 2 cm (0.8 in) wide. The two central leaflets were elliptical and symmetrical, and the two outer leaflets were asymmetrical, with the outside half of the leaflet considerably wider. Leaflets had a strong central vein, with smaller veins diverging, dividing, and rejoining toward the leaflet margin. The foliage was found with three-part bundles of pollen sacs assigned to *Caytonanthus* and isolated *Caytonia* "fruits" and branches from which the individual seed-bearing structures had fallen.

FOUND IN Sandstones and volcanic ash of the Lonco Trapial Formation.

HABITAT Lakeside forest.

NOTES The initial diversity of the Caytoniales was described only from the Northern Hemisphere. This species demonstrates that the group had a worldwide distribution.

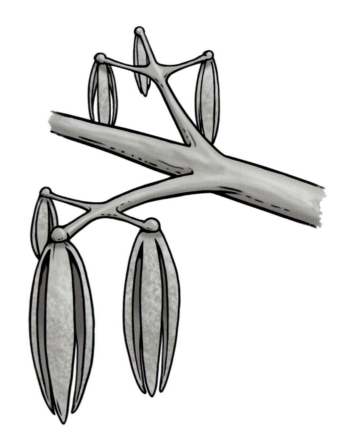

Caytonanthus pollen organ from Patagonia.

CORYSTOSPERMALES

The Corystospermales are sometimes grouped with the Caytoniales and Peltaspermales into an informal group called "seed ferns"—plants with fernlike foliage and seeds. The Corystospermales were a relatively small group of woody shrubs or small trees found mostly in the southern continents of Gondwana, although a few have been reported from the Triassic of China and the Jurassic of Europe. *Dicroidium* was a ubiquitous leaf of Gondwana in the Mesozoic, and a succession of species define the ages of the Triassic in this region.
MORPHOLOGICAL FEATURES Woody trees with short and long shoot branches much like ginkgos. Leaves had a sturdy petiole divided into two or three branches that displayed fernlike pinnae with dichotomously branching veins. Many foliage species have been described across the group's spatial and temporal range. Seed and pollen organs developed on short shoots.

Dicroidium odontopteroides, Umkomasia macleanii, and *Pteruchus africanus*

LOCATION Umkomaas Valley, KwaZulu-Natal, South Africa.
AGE Late Triassic, Carnian (237–227 million years ago).

CHARACTERISTICS Leaves with a sturdy petiole about 2 mm (0.08 in) wide that divided once. Pinnules with rounded triangular shape alternated on either side of the leaf midrib. Smaller veins branched dichotomously. Wood of the form *Rhexoxylon tetrapteridoides* resembled conifer wood but divided into wedges rather than produced in a continuous ring. Seed-bearing *Umkomasia macleanii* composed of

Dicroidium plant with *Pteruchus* pollen organs (upper right) and *Umkomasia* seed-bearing structures

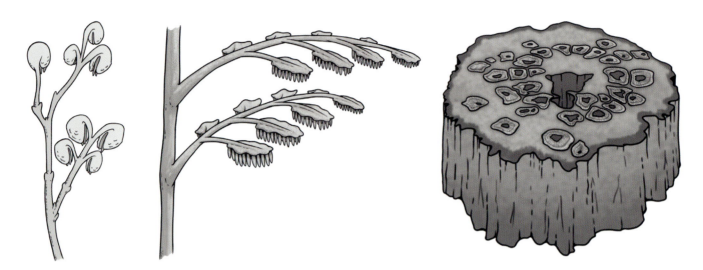

LEFT: *Umkomasia macleanii* seed-bearing structures; MIDDLE: *Pteruchus africanus* pollen organ; RIGHT: *Rhexoxylon* wood of *Dicroidium*.

a branch up to 5 cm (2 in) long, with lateral branches up to 1 cm (0.4 in) long upon which pairs of ovules developed. Each ovule hung from a small cup-shaped structure attached to the branch by a small stalk. Ovules had extensions several millimeters long that may have produced a drop of liquid to help trap wind-wafted pollen. Pollen-bearing structures, *Pteruchus africanus*, were supported on lateral branches up to 6 cm (2.4 in) long. Lateral branches emerged on either side to support a single leaflike structure from which

many pollen sacs hung. *Pteruchus africanus* produced pollen with two air-filled bladders that made the large grains buoyant on the wind.

FOUND IN Organic rich shales of the Molteno Formation.

HABITAT Floodplains with waterlogged soils.

NOTES *Umkomasia* branches are sometimes found without the associated seeds, suggesting that the seeds fell off—or were taken—at maturity rather than being dispersed as a unit.

GNETALES

The living Gnetales includes three genera: *Gnetum*, *Ephedra*, and *Welwitschia*. *Gnetum* includes about 50 living species of tropical shrubs and vines with wood containing large conducting cells, much like those of flowering plants, and broad, net-veined leaves. *Ephedra* includes about 60 living species of bushy shrubs with green photosynthetic stems. Most species produce seeds and pollen on separate individuals. *Welwitschia* includes a single species, *Welwitschia mirabilis*, found only in the Namib Desert of southern Africa. Plants produce a single pair of ribbon-like leaves that grow continually throughout a plant's life. Genetic studies show that the three genera are closely related despite their very different appearances.

The fossil record of the Gnetales is sparce. If living species are a guide, a preference for dry environments and tropical forests where decomposition is rapid may explain the paucity of fossils.

Gnetum gnemon.

Ephedra viridis.

Welwitschia mirabilis.

RIGHT: Pleated pollen (20–80 μm in length) typical of Mesozoic members of the Gnetales.

MORPHOLOGICAL FEATURES The three living members of the group share few features. *Ephedra* and *Welwitschia* have pleated pollen, which is the fossil signature of the group throughout most of its history. In contrast, *Gnetum* pollen resembles spiky spheres.

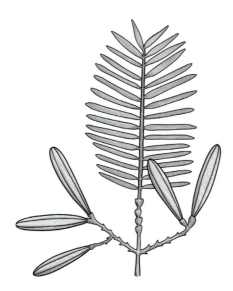

Dechellyia gormanii with seed-bearing branches.

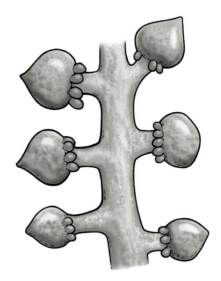

Masculostrobus clathratus pollen organs.

Dechellyia gormanii, *Masculostrobus clathratus,* and *Equisetosporites chinleana*

LOCATION Canyon de Chelly, northeastern Arizona, USA.

AGE Late Triassic, Norian to Rhaetian (227–201 million years ago).

CHARACTERISTICS Branches 2–5 mm (0.08–0.2 in) wide bore leaves 2–10 cm (0.8–3.9 in) long and 1–2 mm (0.04–0.2 in) wide arranged opposite one another. Small scale leaves pressed tightly to the branch near the main stem. Each larger leaf had a strong central vein with delicate, parallel veins on either side. Branches with ovules were covered in scale leaves and ended in a pair of leaves that folded and fused at their bases into a container for a single seed, 7 mm (0.3 in) long and 3 mm (0.1 in) wide. Above the ovule, the seed-bearing leaves widened to about 1 cm (0.4 in) and extended 3–4 cm (1.2–1.6 in), with a rounded tip and two broad ribs. Pollen cones, *Masculostrobus clathratus*, were about 2 cm (0.8 in) long and 1 cm (0.4 in) wide, composed of a spiral of stalked, triangular appendages bearing clusters of pollen sacs that produced *Equisetosporites chinleana* pleated pollen.

FOUND IN Mudstones of the Chinle Formation.

HABITAT Floodplain forest.

NOTES When Sid Ash described *Dechellyia gormanii*, he called it "a conifer of uncertain affinities" because it appeared to have a conifer-like cone. However, the presence of the typically Gnetalean pleated pollen *Equisetosporites chinleana* within the cone suggested affinities with the Gnetales.

PELTASPERMALES

These seed ferns—seed plants with fernlike foliage—were originally described from the Late Triassic of Greenland and South Africa. Today, fossils confirm a global distribution that began in the Pennsylvanian (as early as 323 million years ago) and extended through the Triassic. The group included a hodgepodge of plants with a wide variety of features that link them to the ginkgos and corystosperm seed ferns.

MORPHOLOGICAL FEATURES Plants of the Peltaspermales made leaves that could be pinnate and fernlike, forked, or simple with smooth margins. Few pollen organs are known, but those that have been described produced pollen sacs on leaves stripped down to the supporting veins. Seed-producing structures unite the group. In all species, ovules developed under an umbrella-like structure, with the umbrella handle attached to a specialized branch.

Peltaspermum rotula, Antevsia zeilleri, and *Lepidopteris ottonis* foliage

LOCATION Scoresby Sound, eastern Greenland.
AGE Late Triassic, Rhaetian (208–201 million years ago).

CHARACTERISTICS *Lepidopteris ottonis* leaves were once-pinnate and had a thick outer covering, suggesting that they were adapted to sunny or dry environments. Stomata were surrounded by distinctive hairs to slow water loss. *Peltaspermum rotula*

LEFT: *Lepidopteris ottonis* foliage with *Antevsia* pollen organs.

ABOVE: *Lepidopteris ottonis* foliage with *Peltaspermum* seed-bearing structures.

FAR LEFT: *Peltaspermum rotula* seed-bearing structure.

LEFT: *Antevsia zeilleri* pollen organ.

seed-bearing organ consisted of about 20 seeds that developed under an umbrella-shaped structure 1–1.2 cm (0.4–0.5 in) in diameter, with lobed margin and blister-like bumps on the surface. Several seed-bearing umbrellas attached to a dangling branch. Pollen organ, *Antevsia zeilleri*, about 10 cm (3.9 in) long, with branches each about 2 cm (0.8 in) long. Branches divided in two and supported one or two rows of oval pollen sacs.

FOUND IN Mudstones of the Kap Stewart Group.

HABITAT Floodplain or coastal environment in a warm, possibly arid, climate.

NOTES *Lepidopteris ottonis* leaves were so common across the Northern Hemisphere in the latest Triassic that it could be used to age date rock layers. Toward the end of the Triassic, *Lepidopteris ottonis* began producing large, abnormal pollen grains, indicating ecological stress associated with rapid, unpredictable climate change ahead of the end-Triassic extinction event. *Lepidopteris ottonis* did not survive the Triassic and its pollen type disappeared as the Jurassic dawned.

Umaltolepis mongoliensis with *Pseudotorellia baganuriana* foliage

LOCATION Tevshiin Govi Coal Mine, central Mongolia.

AGE Early Cretaceous, Aptian to Albian (125–100 million years ago).

CHARACTERISTICS The leaves, *Pseudotorellia baganuriana*, were long and slender with a tapered base, pointed tip, and many fine, parallel veins. They ranged from 40–48 cm (15.7–18.9 in) long and 4–7 cm (1.6–2.8 in) wide, although smaller but otherwise identical leaves were also found. Leaves attached to a branching short shoot covered with scaly bark. The seed-bearing structure, *Umaltolepis mongoliensis*, consisted of an umbrella-shaped structure supported on a resin-infused stalk that also developed on a short shoot. The umbrella consisted of four segments that opened to reveal four ovules hanging beneath. Winged seeds matured under the umbrella, which likely fell apart when the seeds were ready for dispersal. Pollen organs

are unknown, although pollen with two bladders has been found in association with *Umaltolepis mongoliensis*.

FOUND IN Mudstones of the Huhteeg Formation.

HABITAT Marshy floodplain forest or swamp.

NOTES Three different leaf species, *Pseudotorellia baganuriana*, *P. resinosa*, and *P. palustris*, have been found in the same layers as *Umaltolepis mongoliensis*. The species differ most significantly in size and some other details of surface structure. While they could represent different plants, they could also be variations associated with different-age shoots or different light conditions. Since no leaves have yet been found attached to shoots bearing *Umaltolepis mongoliensis*, these hypotheses remain open.

Umaltolepis mongoliensis.

PENTOXYLALES

This group was initially described from Jurassic and Early Cretaceous rocks of the Rajmahal Hills of northeastern India by Birbal Sahni, a paleobotanical pioneer in the region. Subsequently, members of the Pentoxylales have been described from across Gondwana. Their slender woody stems led many to reconstruct the group as vines or small trees.

MORPHOLOGICAL FEATURES The group takes its name from the five wedges of wood that characterize the stem. Leaves were elongate, with a strong central rib and smaller veins emerging from it at right angles. Pollen produced by a tuft of pollen sacs, and ovules were displayed in a cluster, suggesting insect pollination.

Pentoxylon sahnii, Carnoconites compactum, Sahnia nipaniensis, and *Nipaniophyllum raoi* foliage

LOCATION Nipania, in the Rajmahal Hills, Bihar, India.

AGE Late Jurassic, Bajocian to Oxfordian (170–167 million years ago).

CHARACTERISTICS Stems, *Pentoxylon sahnii,* 3 mm–2 cm (0.1–0.8 in) in diameter, representing different parts of a bushy plant. Each had five separate wedge-shaped bundles of wood produced from clusters of wood-producing cells rather than a continuous ring. Clusters of thick leaves called *Nipaniophyllum roai* attached to short shoots with a short petiole. Leaves were up to 20 cm (7.9 in) long and strap-shaped, with a strong midrib and delicate lateral veins that emerged perpendicular to the midrib and extended to the leaf edge. Pollen structures, *Sahnia nipaniensis,* developed at the ends of branches. Many stalks a few centimeters long emerged from a dome-shaped structure. Each stalk bore clusters of pear-shaped modified leaves, each of which held a cluster of pollen sacs. Seed-bearing structures, *Carnoconites compactum,* also developed at branch tips. An irregularly branched stalk developed dozens of ovules that packed together into a head.

FOUND IN Cherts formed when mineral-rich water penetrated peat deposits.

HABITAT Wetlands or lake margins.

Pentoxylon sahnii.

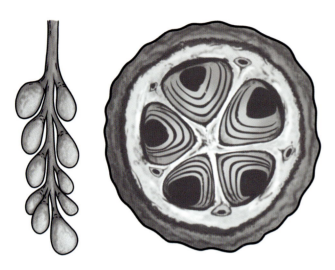

LEFT: *Sahnia nipaniensis* pollen organ.

RIGHT: *Pentoxylon sahnii* stem cross section showing five wood wedges

Nipaniophyllum raoi foliage.

NOTES *Pentoxylon sahnii* was named for its discoverer, Birbal Sahni, who collected the original chert blocks from which the plant was described. Born in what is today the Pakistani Punjab, Sahni studied with Albert Seward at the University of London before returning to India in the 1930s to build a community of Indian paleobotanists. Although a strong believer in the importance of international friendships and collaboration, Sahni spent his short career training Indian paleobotanists, eventually founding a research institute that today bears his name—the only one of its kind to focus exclusively on paleobotany.

Carnoconites compactum seed-bearing structures.

SEED PLANTS WITH UNCERTAIN RELATIONSHIPS

These profiled do not fit comfortably into any group. They have been proposed to be ancestors of, or perhaps even early members of, the flowering plants. However, imperfect preservation leaves their relationships mysterious.

Furcula granulifera

LOCATION Scoresby Sound, eastern Greenland.
AGE Late Triassic, Rhaetian (208–201 million years ago).
CHARACTERISTICS Elongated leaves 7–15 cm (2.8–5.9 in) long and 6–18 mm (0.2–0.7 in) wide, with pointed tips and smooth margins. Many forked in the middle. Each branch had a prominent midrib with smaller lateral veins diverging at 2 mm (0.08 in) intervals. Each vein forked once before reaching the margin. Between these veins, still smaller veins developed a fine network. The leaf was thin and delicate, with stomata on the lower surface lining the parallel veins. No reproductive structures have been found.
FOUND IN Organic rich shales of the Kap Stewart Group.

HABITAT Floodplain wetland or coastal marsh environment in a warm and dry climate.
NOTES *Furcula granulifer* is a rare and poorly known plant. About 50 specimens were collected during the 1926–27 Danish expedition to eastern Greenland. However, its vein network caused some stir in the search for pre-Cretaceous flowering plants. Tom Harris raised a scholarly eyebrow in his 1932 description, noting that its venation resembled living angiosperms and some ferns. He wrote, "if *Furcula* were a Tertiary fossil . . . it would unquestionably be regarded as [an angiosperm] on the evidence of its venation alone. As, however, it is so much older than any other fossil [angiosperm], I prefer to consider the question of its classification open."

Pannaulika triassica

LOCATION Solite Quarry, Dan River Basin, southern Virginia, USA.
AGE Late Triassic, Carnian (237–227 million years ago).
CHARACTERISTICS Leaf fragment, possibly one part of a larger, three-lobed leaf, approximately 3 cm (1.2 in) long, with a prominent central vein and smaller veins branching from it on alternate sides of the central vein. Veins approached the margin of the leaf and then curved toward the leaf tip to connect with the adjacent vein to form a loop. Dense polygonal network of smaller veins filled in around the loops. A poorly preserved reproductive structure was found nearby but not in direct association with *Pannaulika triassica*. It consisted of a stalk about 5 cm (2 in) long to which were attached a series of bracts, each surrounding an ovule. Similar structures were found as dispersed seeds.

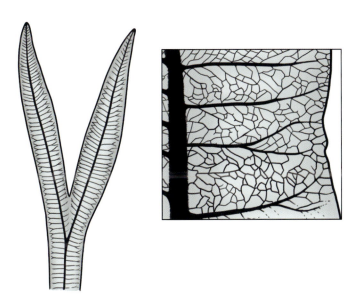

Furcula granulifera leaf with net veins.

Pannaulika triassica leaf with net veins.

Pannaulika triassica seed head.

FOUND IN Black mudstones of the Cow Branch Formation.

HABITAT Margin of a quiet pond.

NOTES In describing *Pannaulika*, Bruce Cornet used jargon usually restricted to angiosperms to bolster his case for flowering-plant affinities. The discovery of a Triassic or Jurassic flowering plant would be extraordinary, and such a claim would require overwhelming evidence. Unfortunately, this scrap of leaf, although tantalizing, does not provide sufficient evidence to extend the angiosperm fossil record into the Triassic.

Sanmiguelia lewisii, Axelrodia burgerii, and Synangispadixis tidwellii

LOCATION San Miguel River valley, southwestern Colorado, USA. Also found in west Texas.

AGE Late Triassic, Norian to Rhaetian (227–201 million years ago).

CHARACTERISTICS Woody stems that produced a spiral of large elliptic leaves up to 40 cm (15.7 in)

Sanmiguelia lewisii.

long and 25 cm (9.8 in) wide. Leaves were distinctly pleated, with parallel veins. Pollen organs called *Synangispadixis tidwellii* consisted of a spike about 1.5 cm (0.6 in) long, covered in minute leaflike structures, each bearing a pair of pollen sacs. Seed-producing structures, *Axelrodia burgerii*, developed in clusters on short branches. Each ovule-bearing structure was surrounded by a whorl of hairy bracts. Ovules surrounded by tissue that was open at one end to receive pollen.

FOUND IN Red sandstones of the Delores Formation.

HABITAT River and coastal lowlands in an arid climate.

NOTES When Roland Brown first reported *Sanmiguelia* leaves in 1956, he confidently called it a palm. Thirty years later, Bruce Cornet discovered more fossils in similar-age rocks in west Texas. This time, pollen and seed organs were preserved near the leaves. Cornet concluded that the plant had angiosperm-like ovules wrapped in two layers of tissue that developed into fruit. However, these interpretations were limited by preservation. Unlike angiosperms, *Sanmiguelia* produced separate pollen and seed organs and had conifer-like wood and cycad-like pollen, all of which appear to take it out of contention for the earliest flowering plant.

Synangispadixis tidwellii pollen organ.

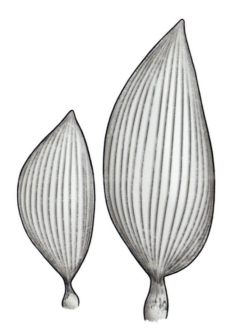

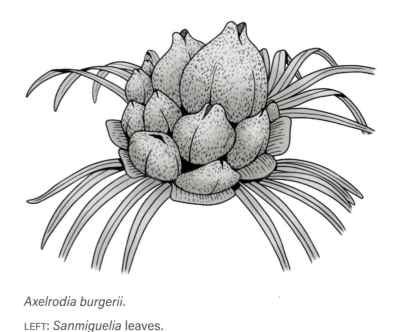

Axelrodia burgerii.

LEFT: *Sanmiguelia* leaves.

FLOWERING PLANTS

Pollen from the Early Cretaceous (Valanginian to early Hauterivian, 140–131 million years ago), with a characteristic multilayered wall and intertwining surface ridges, is the oldest undisputed fossil evidence of flowering plants. This timing agrees with genetic studies that place angiosperm origins between the end of the Early Triassic (247 million years ago) and the middle of the Late Jurassic (154 million years ago). This research also suggests a period of rapid diversification in the Late Jurassic. The time gap between their origin and the first fossil record suggests that flowering plants began their evolutionary history in an environment where fossils were unlikely to form.

Today, angiosperms include more than 300,000 species that inhabit almost all land and freshwater habitats. Such evolutionary success relies on a suite of features that, together, open both ecological and evolutionary doors. While many Mesozoic plants evolved some of these features, only the angiosperms assembled the complete package.

Like all plants, angiosperms tend to fall apart before they become fossils. Flowering plants also favor chemical defenses for their leaves and produce delicate reproductive structures—flowers—that render these parts unlikely to preserve. The tendency to decay and disperse means that angiosperm parts are seldom found together in the fossil record. Therefore, most profiled species feature just one plant part.

MORPHOLOGICAL FEATURES The ephemeral angiosperm flower is the group's most conspicuous feature. Flowers include ovules reduced to just a handful of cells and completely encased by two layers of protective tissue. For fertilization, a tube must grow from the pollen grain through the parental tissue to deliver sperm to the egg cell. This allows

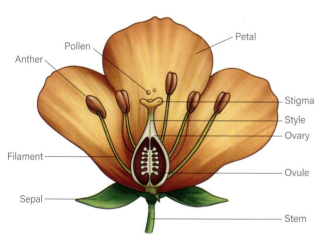

Parts of a flower.

The features of some flowering plant leaves.

OPPOSITE: Aquatic *Archaefructus liaoningensis* in its pond habitat.

the parent plant to control which pollen grain fertilizes the egg cell, providing a boost to diversification as mate choice does in animals. Flowers commonly have a ring of pollen-producing structures—anthers—surrounding the egg-producing parts, and these are surrounded by modified leaves—petals and sepals—that, in many species, help attract pollinators.

Specialized relationships boost angiosperm diversification potential. Evolution honed many angiosperm flowers to attract a few faithful pollinator species. Fertilized ovules then develop into seeds surrounded by nutrient-packed treats—fruit—for potential seed dispersers. Although not unique to flowering plants, angiosperms refined pollinator and disperser relationships, which bolstered diversification.

Other features gave angiosperms an ecological edge. Many flowering plants have enlarged water-conducting cells that help move water efficiently to thirsty leaves. This helps boost photosynthetic rate. Angiosperms also possess cellular and molecular features that allow them to convert light energy to sugar faster and respond to changes in their environment more quickly than other plants.

EARLY CRETACEOUS ANGIOSPERMS

Spherical pollen grains from the Negev Desert

LOCATION Kokhave 2 borehole, Ashkelon, Israel.
AGE Early Cretaceous, late Valanginian to early Hauterivian (134–131 million years ago).
CHARACTERISTICS Small, spherical pollen grains ranging from 15–26 µm in diameter, with the multilayer wall structure characteristic of all flowering plants. The outermost layer has an irregular network of thickenings but no obvious openings.
FOUND IN Mudstones of the Helez Formation.

HABITAT Tropical coastal plain or estuary.
NOTES These pollen grains are the oldest accepted flowering plant fossils. In these oldest sediments, angiosperm pollen made up fewer than 2 grains per 1,000 examined. Altogether, only 15 grains have been found—too few for formal naming of the pollen species.

Acaciaephyllum spatulatum

LOCATION Richmond, Virginia, USA.
AGE Early Cretaceous, Aptian (125–113 million years ago).
CHARACTERISTICS Elliptical leaves 2–4 cm (0.8–1.6 in) long and about 1 cm (0.4 in) wide.

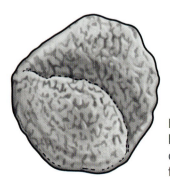

Early Cretaceous pollen believed to be the first definitive fossil evidence of flowering plants.

Acaciaephyllum spatulatum.

Leaves attached to a slender stem with a 2–3 mm (0.08–0.1 in) wide petiole that continued to widen into the blade of the leaf. Two to three veins entered the leaf at its base and divided to fill in the surface of the leaf as it widened. Smaller veins filled in the leaf surface with a haphazard network.

FOUND IN Sandstones of the Patuxent Formation.

HABITAT Margins of a fast-flowing stream or river. Many specimens were found on the Cretaceous stream banks, suggesting that the species grew quickly and completed its life cycle between flood events.

NOTES Although the name hints at a connection with living *Acacia*, the fine-vein network of *Acaciaephyllum spatulatum* is poorly organized and irregular compared to modern *Acacia*.

Afropollis jardinus

LOCATION M'Bour borehole, Senegal.

AGE Early Cretaceous, Albian (113–100 million years ago).

CHARACTERISTICS Small, spherical pollen grains with the multilayer wall found only in flowering plants. They lacked a well-defined opening and were covered in an open network of bumpy ridges.

FOUND IN Mudstones of the N'Toum Formation.

HABITAT Pollen grains preserved in marine mudstones that yield little information about the environment in which the plants grew.

NOTES *Afropollis jardinus* is known from the Barremian (129–125 million years ago) of Nova Scotia, England, Morocco, Libya, and Israel, and from the Aptian (125–113 million years ago) of Gabon, Egypt, Peru, and the eastern USA. Whatever plants produced it were widespread across the tropical Early Cretaceous. However, not all *Afropollis*

species possessed all the features characteristic of angiosperms. *Afropollis* species reported from the Triassic and Jurassic may have been produced by an extinct angiosperm ancestor that lived in tropical Gondwana.

Araliaephyllum westonii

LOCATION Mill Creek, southwestern Alberta, Canada.

AGE Early Cretaceous, Albian (113–100 million years ago).

CHARACTERISTICS Leaves 6–12 cm (2.4–4.7 in) wide, with three to five triangular lobes. A narrow petiole connected the leaf to a woody parent plant. Three prominent veins diverged from the petiole and extended into the central three lobes. In five-lobed examples, smaller veins diverged from the outer two main veins to enter these lobes. Within each lobe, smaller veins created an irregular network. Large leaves had a flat base, while smaller forms were pointed.

Afropollis jardinus pollen.

Araliaephyllum westonii.

FOUND IN Volcanic ash of the Crowsnest Formation.
HABITAT Floodplain woodland with a variety of flowering trees or shrubs and an understory of ferns.
NOTES The rocks of the Crowsnest Formation were deposited when volcanoes to the west exploded and sent fiery ash rolling down the mountainsides, smothering the forests on the mountain's flanks.

Araripia florifera

LOCATION Araripe Plateau, northeastern Brazil.
AGE Early Cretaceous, Aptian (125–113 million years ago).
CHARACTERISTICS Woody twigs with flowers and a variety of leaf forms. Most leaves were 2–3 cm (0.8–1.2 in) long and irregularly lobed. A prominent central vein extended from the petiole to the tip, which was rounded, with the vein extending past the leaf margin into a small spine. Smaller veins branched from the central vein and extended into other lobes, ending in spines. Veins that did not enter lobes curved toward the leaf tip as they approached the margin. Flowers developed on a short stalk between the branch and a leaf. Flower buds 1 cm (0.4 in) long, with narrow petals arranged in a spiral around the flower's core. A spatula-shaped bract supported the flower. Mature flowers shed their petals, leaving the exposed ovule-bearing structure. Pollen-bearing structures were not preserved so we do not know whether this plant produced seeds and pollen on the same flower.
FOUND IN An 8 m (26 ft) thick section of the Crato Formation known as the Crato Konservat Lagerstätte, a zone of extraordinary fossil preservation.
HABITAT Edge of a brackish lagoon in a semiarid climate.
NOTES Evolutionary analysis identifies *Araripia florifera* as an early branch of the laurel family.

Archaefructus eoflora

LOCATION Beipiao, western Liaoning Province, China.
AGE Early Cretaceous, Barremian–Aptian boundary (radioisotopically dated to 125 million years ago).
CHARACTERISTICS The whole plant was 27 cm (10.6 in) high, with a horizontal branching stem that had delicate roots. Feathery leaves had limited waterproofing on their surfaces, suggesting that they were submerged in water. Long reproductive shoots emerged from the horizontal stem and produced a spiral of petals that surrounded a spiral of anthers that encircled a single ovule-bearing structure. Mature fruits 1.8 cm (0.7 in) long produced five to eight seeds.
FOUND IN Muddy shales of the Yixian Formation.

Araripia florifera.

Archaefructus eoflora.

HABITAT Aquatic— shallows of a large lake. **NOTES** *Archaefructus* is a paleobotanical all-star for being preserved as a whole plant, including roots, stem, leaves, flowers, and fruit. The plants grew with their roots and horizontal stems buried in the muddy lake bottom, their feathery leaves supported by water and flowers that floated on or extended above the surface. *Archaefructus eoflora* flowers produced pollen *and* seeds, a feature that links the species with living water lilies.

Archaefructus liaoningensis

LOCATION Beipiao, western Liaoning Province, China.
AGE Early Cretaceous, Barremian–Aptian boundary (approximately 125 million years ago).
CHARACTERISTICS Branching stems bore fruit arranged in a spiral. Fruit attached with a short stem and contained two to four seeds. Flowers lacked petals, with only anthers surrounding a single ovule-bearing structure. Leaves were small and feathery with a thin waterproof covering, suggesting that they grew submerged.

Archaefructus liaoningensis.

FOUND IN Muddy shales of the Yixian Formation.
HABITAT Aquatic—shallows of a large lake.
NOTES When *Archaefructus liaoningensis* was first described, it was assigned a Jurassic age and hailed the oldest flowering plant, but the Jurassic age was speculative. A year later, radioisotopic dating confirmed that the fossils were 125 million years old—Early Cretaceous. The age update did not make *Archaefructus* any less spectacular—flowers are seldom so beautifully preserved.

Seed heads of
Archaefructus liaoningensis.

Canrightia resinifera

LOCATION Catefica, Portugal.

AGE Early Cretaceous, late Barremian to Aptian (126–113 million years ago).

CHARACTERISTICS Flowers and fruit were almost circular, with the flowers containing two or five upright ovules completely enclosed by parental tissue and packed with small, spherical resin bodies that may have discouraged predators. Anthers surrounded the central part of the flower, which were, in turn, surrounded by a small cup of fused petals.

FOUND IN River sands and clay of the Almargem Formation.

HABITAT Banks of a fast-flowing river.

NOTES Hundreds of specimens of this species have been recovered from Early Cretaceous sediments of Portugal. Some were preserved as charcoal, indicating that the flowers burned in a wildfire before dropping into a shallow backwater to be buried in clay. Others dropped fresh from their parent and were buried. *Canrightia resinifera* does not fit into any modern angiosperm group.

Central part of the *Canrightia resinifera* flower.

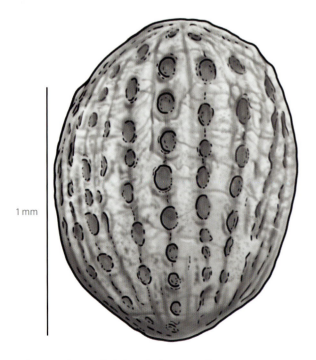

The seed *Canrightiopsis intermedia*.

Canrightiopsis intermedia

LOCATION Lusitanian Basin, western Portugal.

AGE Early Cretaceous, Aptian (125–113 million years ago).

CHARACTERISTICS Small, elliptical fruit 1.1–1.3 mm (0.04–0.05 in) long and about 1 mm (0.04 in) wide. Thick inner layer filled with oil bodies.

FOUND IN Clays of the Calvaria Member of the Figueira da Foz Formation.

HABITAT Open streamside.

NOTES This fruit bridges the gap between the common fossil species *Canrightia resinifera* and the angiosperm family that includes living *Chloranthus*.

Ficophyllum crassinerve

LOCATION Fredericksburg, Virginia, USA.

AGE Early Cretaceous, Aptian or earliest Albian (125–113 million years ago).

CHARACTERISTICS Leaves at least 15 cm (5.9 in) long and 8 cm (3.1 in) wide at the base. Strong central vein up to 5 mm (0.2 in) wide at the point where it attached to the petiole and tapered toward the leaf tip. Both the

Two fossil specimens united under the name *Ficophyllum crassinerve.*

base and tip of the leaves were pointed. Smaller veins emerged from the central vein at irregular intervals and extended toward the leaf's smooth margin before turning toward the tip and forming a loop with the vein above. Still smaller veins filled in an irregular network.

FOUND IN Organic-rich sediments of the Arundel Clay.

HABITAT Lakeshore marsh.

NOTES When William Morris Fontaine collected the first specimens of this species, he found them in broken pieces and mistakenly assigned the top (*Ficophyllum crassinerve*) and bottom (*Proteaephyllum ellipticum*) halves of the leaf to separate species. When Leo J. Hickey took a second look, he recognized the two parts of the same leaf and reunited them. Technically, the reunited form should be called *Proteaephyllum ellipticum* because this species was described on page 285 of Fontaine's 1889 monograph and *Ficophyllum crassinerve* appeared on page 292. However, Hickey noted that the leaf's venation corresponded best to the genus *Ficophyllum.*

Ficophyllum palustris

LOCATION Alexander Island, Antarctica.

AGE Early Cretaceous, Albian (113–100 million years ago).

CHARACTERISTICS Pear-shaped leaves with a rounded bottom and elongated, pointy tip. Leaves up to 7 cm (2.8 in) long and 5 cm (2 in) wide. Margin was serrated. A prominent central vein ran from the petiole to the tip. From this vein, smaller veins branched on opposite sides of the leaf at roughly right angles, continued toward the leaf margin, then curved

Ficophyllum palustris.

toward the leaf tip to make a loop. Smaller veins filled in an irregular network.

FOUND IN Organic-rich siltstones of the Triton Point Member of the Neptune Glacier Formation.

HABITAT Swamps and marshy areas of the floodplain.

NOTES The name *Ficophyllum* captures leaves that are longer than wide, with looping venation. Although such leaves resemble those of several modern angiosperm groups, without flowers or fruit they cannot be confidently assigned to any group.

Gnafalea jeffersonii

LOCATION Alexander Island, Antarctica.

AGE Early Cretaceous, Albian (113–100 million years ago).

CHARACTERISTICS Elliptical leaves 1.5–4.7 cm (0.6–1.9 in) long and 1–2.8 cm (0.4–1.1 in) wide, with a pointed tip, rounded base, and short petiole. Leaf margin had large teeth with fine serration. A single prominent vein extended the length of the leaf, with smaller veins branching from it in pairs. Each of these veins ended at the leaf margin in a small gland. Smaller veins filled the leaf with an irregular network.

Gnafalea jeffersonii.

Antarctica in the Albian.

FOUND IN Fine-grained sandstones and siltstones of the Triton Point Member of the Neptune Glacier Formation.
HABITAT Floodplain close to the river's edge.
NOTES *Gnafalea jeffersonii* was found with the leaves of several other angiosperms and ferns, suggesting a shrubby thicket growing at the boundary between the frequently flooded river margin and stable, forested ground.

Hydrocotylophyllum alexandri

LOCATION Alexander Island, Antarctica.
AGE Early Cretaceous, Albian (113–100 million years ago).
CHARACTERISTICS Small leaves 1.5–3.5 cm (0.6–1.4 in) long and 2.5–3 cm (1–1.2 in) wide, with a rounded tip, asymmetrical base, and short petiole. A series of veins radiated from the petiole and spread out across the leaf, branching several times as they approached the margin. Between these main veins, smaller veins filled in with an irregular network. Leaf margin was wavy with small glands.

FOUND IN Sandstones and siltstones of the Triton Point Member of the Neptune Glacier Formation.
HABITAT Recently flooded riverside floodplain.
NOTES *Hydrocotylophyllum alexandri* fossils were found in mats in layers above slightly older layers that contained a diverse flora of ginkgos, ferns, and Bennettitales. *Hydrocotylophyllum alexandri* appears to have taken over the river margin after floods buried the preexisting vegetation, highlighting the angiosperm ability to occupy disturbed habitats.

Monetianthus mirus

LOCATION Vale de Água clay pit, western Portugal.
AGE Early Cretaceous, Barremian to Aptian (129–113 million years ago).
CHARACTERISTICS Flower about 3 mm (0.1 in) long and 2 mm (0.08 in) in diameter, radially symmetrical, with a central core surrounded by 12 ovule-bearing structures. These were surrounded by 20 anthers on short filaments, which were encircled by petals that were not preserved.
FOUND IN Clays of the Figueira da Foz Formation.

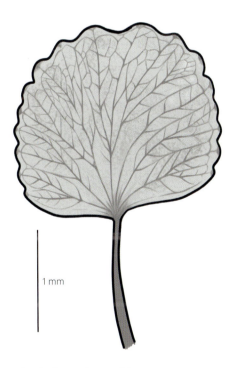

Hydrocotylophyllum alexandri.

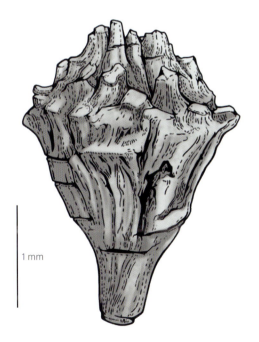

The fossil remains of the tiny *Monetianthus mirus* flower.

187

Monetianthus mirus as it looked in life.

HABITAT Preserved in pond clay.
NOTES This species is known from a single small
flower with features that clearly link it to the living
water lilies. Preservation in pond sediments further
hints at an aquatic lifestyle.

Platanus latiloba

LOCATION Chuwanten Creek, British Columbia,
Canada.
AGE Early Cretaceous, Albian (113–100 million years
ago).
CHARACTERISTICS Large leaves up to 20 cm (7.9 in)
wide, oval, with three to five lobes. Three prominent
veins emerged from the petiole and entered the
three central lobes. In five-lobed leaves, a smaller
vein diverged from the left and right main veins and
entered the lateral lobes. Smaller veins diverged from
the largest veins and continued to divide at irregular
intervals toward the edge of the leaf. Lobes were
rounded with serrated edges. Space between the veins
was filled in with an irregular network of smaller veins.
FOUND IN Winthrop Formation.
HABITAT Diverse rainforest with a conifer and ginkgo
canopy and an understory that included cycads, ferns,
and a variety of other seed plants.

Platanus latiloba.

NOTES Detailed study of the Winthrop
Formation angiosperms showed that
their leaves were thin with relatively
short life spans compared to the other
plants that grew beside them. This
suggests that flowering plants lived a
weedy lifestyle in which they grew and
reproduced quickly, finding space in a
crowded world by favoring disturbed
sites that could not be inhabited by
slow-growing plants.

Rogersia angustifolia

LOCATION Fredericksburg, Virginia,
USA.
AGE Early Cretaceous, Aptian or
earliest Albian (125–113 million years
ago).
CHARACTERISTICS Long, slender leaf
at least 8 cm (3.1 in) long and 1.5 cm

Rogersia angustifolia.

(0.6 in) wide at its widest point. Both base and tip were pointed. A strong central vein formed the petiole and then narrowed as it continued to the leaf tip. Smaller veins emerged from the central vein and divided, eventually looping back on their neighbors near the margin, which was smooth.
FOUND IN Lake sediments of the Arundel Clay.
HABITAT Lakeshore.
NOTES Slender leaves like *Rogersia* were common throughout the Cretaceous. Some early workers linked forms that had serrated margins with the modern willows (*Salix*). However, details of the venation seldom correspond to any living plants, and Aptian *Rogersia* had particularly irregular vein patterns.

Swamyflora alata

LOCATION Puddledock locality, George County, Virginia, USA.
AGE Early Cretaceous, Albian (113–100 million years ago).
CHARACTERISTICS Small flowers, 1.3–2.2 mm (0.05–0.09 in) long and 1.2–1.7 mm (0.05–0.07 in) wide, lacking anthers, and triangular in cross section, with the corners of the triangle extended to small wings. Ovule-bearing

structure contained a single upright ovule. Three fused petals spread out from the top of the ovule-bearing structure, alternating with wings. A slender extension rose from the center of the petals to receive pollen.
FOUND IN Clays associated with river gravels of the Potomac Formation.
HABITAT Riverbank with a forested floodplain.
NOTES The Puddledock locality has yielded a variety of conifer and fern leaves, along with the seeds of Bennettitales and Gnetales, suggesting a diverse forested ecosystem with angiosperms occupying frequently disturbed riverbanks.

Vitiphyllum multifidum

LOCATION Baltimore, Maryland, USA.
AGE Early Cretaceous, Aptian or earliest Albian (125–113 million years ago).
CHARACTERISTICS Leaves at least 6 cm (2.4 in) long and 7 cm (2.8 in) wide, with a strong central vein and pairs of sub-opposite veins that branched from the central vein. Smaller veins branched from these to fill in the surface of the leaf. The leaf had at least three lobes and an irregularly serrated margin.

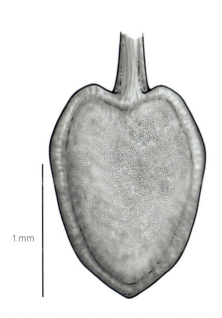

The central part of the flower *Swamyflora alata*.

Vitiphyllum multifidum.

FOUND IN Clay and silt of the Arundel Clay.
HABITAT Lakeshore.
NOTES This species takes its name from a superficial resemblance to grape (*Vitis*) leaves. This Early Cretaceous form has a disorganized array of small veins different from living *Vitis* or even some Late Cretaceous leaves that share the name.

Wasmyflora portugallica

LOCATION Vale de Água clay pit, western Portugal.
AGE Early Cretaceous, Barremian to Aptian (129–113 million years ago).
CHARACTERISTICS Flower 1.7 mm (0.06 in) long and 1.3 mm (0.05 in) wide, without anthers. Seed-producing part contained a single upright ovule that developed below a ring of petals and was surrounded by parental tissue. Fruits were flattened with indistinct wings on either side.
FOUND IN Clays of the Figueira da Foz Formation.
HABITAT Streamside.

1 mm

The fruit of *Wasmyflora portugallica*.

NOTES This flower resembles living *Hedyosmum*, which is among the earliest derived of living flowering plant lineages. *Hedyosmum* produces *Afropollis*–type pollen, which also links it with some of the earliest records of flowering plants in northern Gondwana.

Phytoliths from the tooth residue of a hadrosaur dinosaur

LOCATION Mazongshan area, Gansu Province, northwestern China.
AGE Early Cretaceous, Albian (113–100 million years ago).
CHARACTERISTICS Phytoliths are microscopic silica bodies that form in the cells of many plants and are similar in composition to opal. These hard particles helped deter predators and reinforced tissues. Early grasses produced distinctive phytoliths, including equidimensional forms with curved ends.
FOUND IN The tooth residue of the duck-billed dinosaur *Equijubus normani* preserved in the Zhonggou Formation.
HABITAT A woodland habitat that included conifers, ferns, and very few flowering plants.
NOTES Most of the flowering plant pollen from the Zhonggou Formation came from magnolia-like plants, but evidence of these was not found in the *Equijubus* teeth, suggesting a preference for rare early grasses.

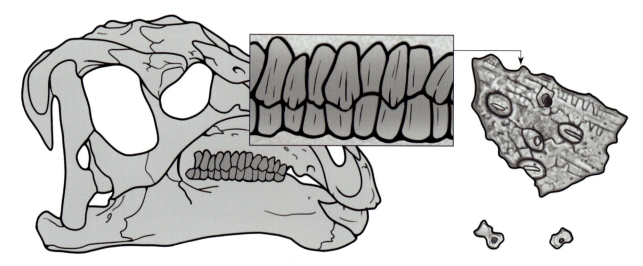

Plant remains, including surface tissue and silica bodies called phytoliths, found on dinosaur teeth show that these animals ate early grasses.

LATE CRETACEOUS ANGIOSPERMS

Archaeanthus linnenbergeri

LOCATION Bunker Hill, Russell County, Kansas, USA.

AGE Mid-Cretaceous, Albian to Cenomanian (113–94 million years ago).

CHARACTERISTICS Small tree or shrub. *Archaeanthus linnenbergeri* seed pods matured in a spiral pattern at the ends of leafy branches. The central core supporting the pods was about 14 cm (5.5 in) long and tapered toward the tip. Immature specimens showed that the core elongated as seed pods matured. Each pod attached to this core with a short stalk that left a diamond-shaped scar. Below the zone where pods developed, the branch bore scars where flower parts attached. Individual pods 2.5–3.5 cm (1–1.4 in) long and 4–7 mm (0.2–0.3 in) wide. Mature pods had a thickened ridge on top to facilitate opening to disperse 10 to 18 seeds each. Seeds were elliptical, 2–3 mm (0.08–0.1 in) long and 1–2 cm (0.4–0.8 in) wide. Pods matured from bottom to top. *Archaepetala beekeri* and *Archaepetala obscura* were petals that had fallen from the flower. *Archaepetala beekeri* was 7–8 cm (2.8–3.1 in) long and 4 cm (1.6 in) wide, elliptical, with a smooth margin and a prominent attachment point that matched the scars below the *Archaeanthus linnenbergeri* seed head. *Archaepetala obscura* was pear-shaped, 8 cm (3.1 in) long and 5 cm (2 in) wide at the wide end. It had a smooth margin and nine prominent veins that radiated from the attachment point. *Kalymmanthus walkeri* bud scales were 2.6–6.5 cm (1–2.6 in) long, 2.1–5 cm (0.8–2 in) wide, and bilobed, with a smooth margin. Tips of the lobes were rounded and rimmed with a curled lip. A prominent vein extended from the base to the point between the two lobes. Two leaf forms accompanied the reproductive structures *Liriophyllus kansense* and *Liriophyllum populoides*. Both leaves had long petioles and two lobes. The lobes were 9–18.5 cm (3.5–7.3 in) long and 8.5–24 cm (3.3–9.4 in) wide, with a single central vein that divided at the base of the lobes and defined the interior margin of each. Smaller veins developed from the largest veins and extended toward the smooth margin. *Liriophyllus kansense* and *Liriophyllum populoides* differed only in that *L. populoides* had small, wing-like extensions at the base of the petiole.

Archaeanthus linnenbergeri.

FOUND IN Gray-brown clay of the Dakota Formation.

HABITAT Marshy swales on the floodplain side of the river's natural levee.

NOTES *Archaeanthus linnenbergeri* was a magnolia-like tree or shrub with large, showy flowers to attract pollinators. The Dakota Formation clays preserved 15 to 20 other kinds of flowering plant leaves, suggesting a diverse angiosperm flora. However, pollen from the same rocks contains only 25% flowering plants, showing that on more stable ground, conifers, cycads, ginkgos, and Bennettitales still flourished.

Bisonia niemii

LOCATION Marmarth, North Dakota, and Buffalo, South Dakota, USA.

AGE Late Cretaceous, Maastrichtian (72–66 million years ago).

CHARACTERISTICS Oval leaves 3–18 cm (1.2–7.1 in) long and 3–16 cm (1.2–6.3 in) wide, with fragments suggesting that leaves could be much larger. Tip rounded. The base of the leaf commonly had lobes that extended on either side of the petiole. Three prominent veins emerged from the petiole and extended into the lobes. These commonly divided to extend into other lateral lobes. Smaller veins extended toward the smooth margin. Additional veins connected the central and lateral veins in upward-curving ladders.

FOUND IN Sandstones of the Hell Creek Formation.

HABITAT Floodplain woodland. In modern tropical and subtropical forests, very large leaves are common on fast-growing trees that exploit disturbed sites.

NOTES Although it resembles some laurels, *Bisonia niemii* does not fit well into any living plant family.

Chloranthistemon alatus

LOCATION Åsen, Scania, southern Sweden.

AGE Late Cretaceous, late Santonian to early Campanian (about 85–82 million years ago).

CHARACTERISTICS Minute flowers, each less than 0.5 mm (0.02 in) long, arranged in pairs along a flower spike up to 2 cm (0.8 in) long. Small, triangular bracts cradled each flower. Flowers were bilaterally symmetrical, with three anthers attached to the ovule-bearing structure, which contained a single ovule. A winglike flap of tissue wrapped around and partially covered the flower.

FOUND IN Clays of the old kaolin quarry near the city of Åsen.

Bisonia niemii.

1 mm

Chloranthistemon alatus.

HABITAT Coastal wetland. Fossils from this site include a diversity of ferns, lycophytes, and cypress conifers.

NOTES Two species of *Chloranthistemon* have been found at this site. Evolutionary analysis allies them with the living genus *Chloranthus*, but pollen features differ significantly from all living species.

Dewalquea smithii

LOCATION White Bluffs along the Black Warrior River, Alabama, USA.

AGE Late Cretaceous, Turonian (94–90 million years ago).

CHARACTERISTICS Leaves with five leaflets that radiated from a single petiole like the fingers of a hand. Petiole divided into three, with the central leaflet attached to the petiole, which divided again to support two additional leaflets on either side. Central leaflet 10–14 cm (3.9–5.5 in) long and 3 cm (1.2 in) wide. Lateral leaflets 9–12 cm (3.5–4.7 in) long and about 3 cm (1.2 in) wide. Each leaflet had a pointed base and tip, with a prominent central vein that ran the length of the leaflet. From it, smaller veins branched and ended in small teeth on the margin.

FOUND IN Sandstones of the Tuscaloosa Formation.

HABITAT Margins of fast-flowing rivers.

NOTES This species was also found in older rocks (late Albian) of Utah, USA. There, *Dewalquea smithii* grew at the edge of a small pond along with about a dozen other angiosperm species. Fossil pollen shows that conifers, cycads, and a diversity of ferns also grew in the region, but none of these plants were found in the pond sediments, suggesting that flowering plants dominated the pond-side flora.

Dryophyllum subfalcatum

LOCATION Lance Creek, Wyoming, USA.

AGE Late Cretaceous, Maastrichtian (72–66 million years ago).

CHARACTERISTICS Leaves 4–12 cm (1.6–4.7 in) long and 1–2.5 cm (0.4–1 in) wide. Pointed tip and base with a petiole up to 3 cm (1.2 in) long. A prominent central vein emerged from the petiole and continued to the tip. Smaller veins diverged and ended in small teeth at the margin. Smaller veins connected these in ladderlike patterns, and a regular network of small veins filled the leaf surface.

FOUND IN River sandstones of the Lance Formation.

HABITAT Streamside shrubs.

NOTES This leaf type was common across the Western Interior of North America during Maastrichtian time. Its evolutionary relationships are unclear. Leo Lesquereux originally allied *Dryophyllum* with oaks, and their overall shape resembles willows, but the fine veins suggest a relationship with the walnuts.

Dewalquea smithii.

Dryophyllum subfalcatum.

Erlingdorfia montana

LOCATION Dawson County, Montana, USA.
AGE Late Cretaceous, Maastrichtian (72–66 million years ago).
CHARACTERISTICS Leaf composed of three leaflets. The central leaflet had three lobes; the lateral leaflets had two unequal lobes. Central leaflet 9–15 cm (3.5–5.9 in) long and 6–12 cm (2.4–4.7 in) wide. Lateral leaflets 8–11 cm (3.1–4.3 in) long and 4–7 cm (1.6–2.8 in) wide. Some specimens had smooth margins, and others had small teeth. A prominent vein entered the central leaflet and divided in three, with one branch entering each lobe. Smaller veins branched from these and continued toward the margin, where they curved along the margin or ended in small teeth. Lateral leaflets had a prominent vein that emerged from the petiole and extended into the larger lobe.

FOUND IN Sandstones of the Hell Creek Formation.
HABITAT Floodplain woodland.
NOTES The species is known from the last two million years of the Cretaceous period in Wyoming, the Dakotas, and Montana, USA. The oldest examples had larger leaflets with smooth margins, while those living near the end of the Cretaceous had smaller leaflets and toothed margins. Since leaf size and the presence of a toothed margin correlate with climate, this trend may reflect adaptation to a cooling and drying latest-Cretaceous climates.

Ficus ceratops

LOCATION Converse County, Wyoming, USA.
AGE Late Cretaceous, Maastrichtian (72–66 million years ago).

CHARACTERISTICS Pear-shaped fruit 3–4 cm (1.2–1.6 in) long and 2–3 cm (0.8–1.2 in) wide at the base. Outer skin of the fruit was thick, with ribs that converged on the narrow end. The narrow neck is commonly broken, but when entire, it ends in a small opening where the fruit attached to the parent plant.
FOUND IN Fine-grained sandstones of the Lance Formation.
HABITAT Streamside woodland.
NOTES *Ficus ceratops* were commonly preserved as casts after the fruits were buried in sandstone and decayed, and the empty space filled with sand. This style of preservation hints that the fruits grew beside and were dispersed by the river. The identity of the fruits has been debated. An early collector, J. B. Hatcher, thought they were bulbs. Frank Hall Knowlton, who described them in

Erlingdorfia montana.

Lesqueria elocata

LOCATION Ellsworth County, Kansas, USA.

AGE Late Cretaceous, Cenomanian (about 100–94 million years ago).

CHARACTERISTICS Seed head bearing 175 to 250 individual, spirally arranged fruits sitting atop a 1 cm (0.4 in) diameter shoot. Each fruit in the seed head had a small keel on the outward-facing side. The tip of the fruit was elongated to a point. Each fruit contained 10 to 20 seeds arranged in two rows. A spiral of flaps about 2 cm (0.8 in) long and 1.7 cm (0.7 in) wide connected to the stem with diamond-shaped attachments. These became smaller down the stem until they were triangular scales at the point where the seed head attached to the main branch.

FOUND IN Three-dimensional molds in the Dakota Sandstone.

TOP: *Guarea guidonia* in the mahogany family resembles *Ficus ceratops*.

ABOVE: *Ficus ceratops*.

RIGHT: *Carpolithus filiferus*.

1911, likened them to figs, placing them in the genus of living figs (*Ficus*). In 1934, Roland Brown described *Carpolithus filiferus* from Eocene-age (56–34 million years ago) rocks of Colorado, USA. These were like *Ficus ceratops* but were preserved squashed in lake-bed clay. With more details preserved, *Ficus ceratops* emerged as a member of the mahogany family.

Lesqueria elocata.

HABITAT Coastal floodplain along the Western Interior Seaway.

NOTES In 1892, Leo Lesquereux described this plant as a Bennettitalean, *Williamsonia elocata*. In 1984, Peter Crane and David Dilcher reexamined the specimens and concluded that the fossils were flowering plants and created the new name *Lesqueria* to honor Lesquereux's initial work.

Liriodendrites bradacii

LOCATION Slope County, North Dakota, USA.
AGE Late Cretaceous, Maastrichtian (72–66 million years ago).
CHARACTERISTICS Bilobed leaf about 8 cm (3.1 in) long and up to 11 cm (4.3 in) wide. Lobe tips rounded and the base of variable shape. A single prominent vein emerged from the petiole and extended to the point where the lobes connected. Smaller veins emerged from the central vein and branched a few times before reaching the smooth margin.
FOUND IN Sandstones of the Hell Creek Formation.
HABITAT Floodplain woodland.
NOTES This variable leaf species might have been considered several species if all the different forms had not been preserved together. *Liriodendrites bradacii* commonly displayed small, circular chew marks in its margin, suggesting that it was a popular insect treat.

Manchestercarpa vancouverensis

LOCATION Shelter Point, Vancouver Island, British Columbia, Canada.
AGE Late Cretaceous, Campanian (100–94 million years ago).
CHARACTERISTICS Fruit 3.5 mm (0.1 in) long and 4 mm (0.2 in) wide, with a hard, woody inner shell containing up to nine individual seed chambers arranged around a hollow central axis. Each seed chamber enclosed one seed. The hard shell was surrounded by a fleshy layer and leathery skin.
FOUND IN Mineral concretions from the Spray Formation.
HABITAT Coastal forest with conifers and tree ferns.

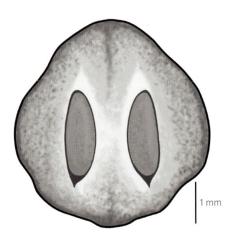

Liriodendrites bradacii.

The fruit *Manchestercarpa vancouverensis*.

NOTES The fossils were found in rocks that also yielded crabs and other marine creatures. The plants that produced *Manchestercarpa vancouverensis* lived in coastal forests or near rivers that carried their fruits to the sea. These fruits belong to the mahogany family.

Marmarthia trivialis

LOCATION Marmarth, North Dakota, USA.
AGE Late Cretaceous, Maastrichtian (72–66 million years ago).
CHARACTERISTICS Leaves elliptical or nearly circular, 4–14 cm (1.6–5.5 in) long and 2–8 cm (0.8–3.1 in) wide. Leaf tip rounded, with three to nine large teeth. Base rounded. Three prominent veins emerged from the petiole and extended into teeth at the leaf margin. Smaller veins branched from the lateral veins and either ended in a large tooth or looped upward along the margin to connect with the vein above. Smaller veins connected the central veins with the two side veins in an irregular, ladderlike pattern.

FOUND IN Sandstones of the Hell Creek Formation.
HABITAT Floodplain or levee woodland.
NOTES This species has been found in Maastrichtian-age rocks of Wyoming, North and South Dakota, Montana, and Colorado, USA. The leaf is common in sandstones associated with floods, suggesting that *Marmarthia trivialis* grew on or very near the river's natural levee.

Microvictoria svitkoana

LOCATION Middlesex County, New Jersey, USA.
AGE Late Cretaceous, Turonian (94–90 million years ago).
CHARACTERISTICS Small flowers, 2–3 mm (0.08–0.1 in) long and 1.2 cm (0.5 in) in diameter, included seed-producing structures surrounded by anthers on a short stalk. Petals developed above the ovule-producing structure.
FOUND IN Old Crossman Clay Pit of the Raritan Formation.
HABITAT Aquatic plant growing in quiet, shallow water.
NOTES The genus takes its name from the modern water lily genus *Victoria* to which the Cretaceous flower was closely related. *Microvictoria* even had a similar pollination system in which its petals snapped shut on visiting beetles, capturing them for a time before releasing them to carry pollen to another flower.

Marmarthia trivialis.

1 mm

Microvictoria svitkoana.

Nothofagus subferruginea

LOCATION King George Island, Antarctica.
AGE Late Cretaceous, Campanian to Maastrichtian (84–66 million years ago).
CHARACTERISTICS Leaves up to 9 cm (3.5 in) long and 5 cm (2 in) wide, with pointed tips and bases. A straight central vein emerged from the petiole and continued to the leaf tip. From this vein, smaller veins emerged as slightly offset pairs and continued straight to the leaf margin, where they ended in irregular teeth.
FOUND IN Volcanic and sedimentary rocks of the Zamek Formation.
HABITAT The forested slopes of volcanic mountains. Petrified conifer wood found with the leaf impressions shows growth rings, indicating a freezing season.
NOTES Today, *Nothofagus* includes more than 40 species distributed throughout South America, Australia, New Zealand, New Guinea, and New Caledonia. *Nothofagus subferruginea* belongs to the living group, meaning that its roots extend deep into the Mesozoic of Gondwana.

Phoenicites imperialis

LOCATION Cumberland, Vancouver Island, British Columbia, Canada.
AGE Late Cretaceous, Coniacian to Santonian (90–84 million years ago).
CHARACTERISTICS Large, pleated leaves at least 1 m (3.3 ft) in length and probably longer. Pleated segments were 8–27 mm (0.3–1.1 in) wide and more than 1 m (3.3 ft) long, attached to a petiole that was about 2.5 cm (1 in) wide. In many specimens, pleated segments remained connected to the edge of the leaf, but in some, they frayed to individual pointed tips.
FOUND IN Shales of the members of the Comox Formation.
HABITAT Swamps at the edge of the floodplain.
NOTES Fossils were found on the spoil piles of No. 8 mine of the Cumberland Coal Company, where they were preserved in the shales that overlay the coal seam. This suggests that the palms grew in the swamps and were killed when a flood buried them in clay.

Nothofagus subferruginea.

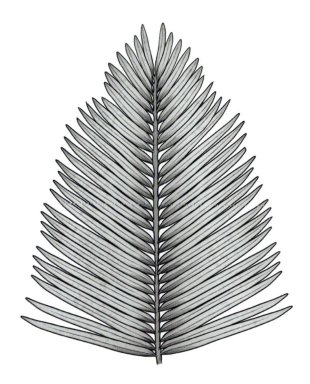

Phoenicites imperialis.

Prisca reynoldsii

LOCATION Hoisington, Kansas, USA.

AGE Late Cretaceous, Cenomanian (100–94 million years ago).

CHARACTERISTICS Woody shrub or small tree with leaves (*Magnoliaephyllum*) that ranged from 10–15 cm (3.9–5.9 in) long and 3–5 cm (1.2–2 in) wide, with a 2 cm (0.8 in) long petiole that flared at the point where it attached to the shoot. Leaves had a prominent central vein with smaller veins extending from it. Lateral veins grew toward the smooth leaf margin and then curved upward to connect with the vein above. Between these, smaller veins created an irregular network. The fruiting structure, *Prisca reynoldsii*, was about 80 cm (2.6 ft) long and hung down on a fibrous central axis with a pointed bract at the point where it attached to the shoot. Individual fruits were arranged in a spiral around the dangling axis. Fruits were elliptical, about 3 mm (0.1 in) long and 2 mm (0.08 in) wide, with a smooth skin.

FOUND IN Sandstones of the Dakota Formation.

HABITAT Streamside woodland.

NOTES The flared ends of *Magnoliaephyllum* petioles suggest that the leaf was shed seasonally.

Prisca reynoldsii.

Sabalites eocenicus

LOCATION Alberta, Canada.

AGE Late Cretaceous, Santonian to Campanian (86–72 million years ago).

CHARACTERISTICS Fan palm leaf, pleated with up to 30 segments, each 1 cm (0.4 in) wide. The central support for the leaf was narrowly triangular with concave sides that extended into a narrow point. This support extended from the petiole, which was about 2.5 cm (1 in) wide and at least 60 cm (2 ft) long. No complete fans were recovered but the largest were at least 45 cm (1.5 ft) long. Pleats separated toward the edge of the leaf and had pointed tips.

FOUND IN Sandstones and shale of the Foremost Formation.

HABITAT Floodplain wetlands.

NOTES This is the northernmost palm reported from the Mesozoic of North America, indicating frost-free subtropical climates during this time.

Sabalites eocenicus.

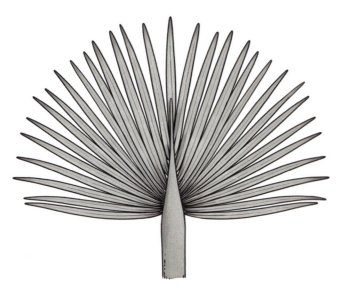

Sabalites eocenicus frond.

Sterculia washburnii

LOCATION Near Tres Lagos, Argentina.
AGE Late Cretaceous, Cenomanian to Coniacian (100–86 million years ago).
CHARACTERISTICS Lobed leaves 2–5 cm (0.8–2 in) long and 3–6 cm (1.2–2.4 in) wide, with smooth margins. Most had three to five lobes. Three major veins emerged from the petiole, with the central vein extending into the lobe at the leaf tip and side veins extending into lateral lobes. In leaves with more than three lobes, the lateral veins branched to send a vein into each additional lobe. Irregularly ladderlike veins connected these major veins, and an irregular network filled in among them.
FOUND IN Mudstones of the Mata Amarilla Formation.
HABITAT Floodplain forest that also included large conifers preserved as upright stumps and a fern understory. Angiosperms were small understory trees or shrubs.
NOTES Edward W. Berry originally described *Sterculia sehuensis* for three-lobed forms and *S. washburnii* for leaves with up to five lobes. Since both forms have now been found together, the collection is better interpreted as a single variable species. Since no flowers or fruit associated with these leaves have been discovered, assigning them to the living genus *Sterculia* is questionable.

Thalassocharis bosquetii

LOCATION Maastricht, the Netherlands.
AGE Late Cretaceous, Maastrichtian (72–66 million years ago).
CHARACTERISTICS Intertwined branching stems with a primary growing tip and side branches. Stems up to 8 mm (0.3 in) in diameter likely grew through the

Sterculia washburnii produced leaves of several different shapes.

sediment and extended leaves into shallow ocean water to form thick carpets. Stems surrounded by a tight spiral of leaf scars. Cross sections of the stem revealed three layers: an outer layer composed of reinforced bundles of water-conducting cells, a middle layer with reinforcing fibers, and an inner layer with air spaces. The center of the stem housed more reinforced water-conducting tubes. Sparsely branched roots emerged from the bases of stems. Leaves of *Thalassotaenia debeyi* were up to 1 m (3.3 ft) long and 3.2 cm (1.3 in) wide, strap-shaped, with smooth, parallel margins. Nine to fifteen veins ran the length of leaves and were connected by cross veins. Air spaces in the stem and open, spongy tissue in the leaves suggest that these plants grew entirely submerged.

FOUND IN Chalk of the Maastricht Formation.

HABITAT Sea stars preserved with the seagrass confirm that it grew in a lagoon.

NOTES At some locations, this seagrass was preserved in thick mats formed when storm waves uprooted the seagrass and washed it shoreward. *Thalassocharis bosquetii* survived the end-Cretaceous extinction event and is found in Cenozoic sediments of the same region.

LEFT: *Thalassocharis bosquetii.*

BELOW LEFT: *Thalassocharis bosquetii* seagrass stem in cross section showing air spaces that kept the plant from drowning.

BELOW: Details of the leaf (left) and leaf attachments (right) reveal close relationships with living seagrasses.

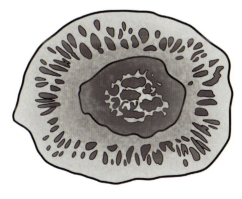

Vitis stantonii

LOCATION Throughout Wyoming and Montana, USA.

AGE Late Cretaceous, Maastrichtian (72–66 million years ago).

CHARACTERISTICS Large, elliptical leaves 15–30 cm (5.9–11.8 in) long and 10–20 cm (3.9–7.9 in) wide. Rounded base and tip. Petiole up to 5 cm (2 in) long, with a flared base, suggesting that the plant shed its leaves regularly. Three prominent veins emerged from the petiole, with smaller veins branching from the two lateral veins. Still smaller veins emerged from these and extended toward the margin, where they ended in irregular teeth. Between the main veins, smaller veins developed a wavy, ladderlike pattern. An irregular network filled in between these veins.

FOUND IN Sandstones of the Lance Formation.

HABITAT Floodplain woodland and riversides.

NOTES Although assigned to the living grape genus *Vitis*, this resemblance is superficial. The species was common all over western North America during Maastrichtian time but became extinct during the end-Cretaceous event.

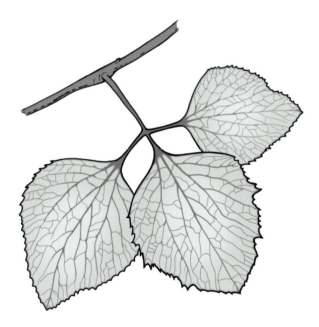

Vitis stantonii.

Phytoliths from sauropod dinosaur coprolites

LOCATION Pisdura, central India.

AGE Late Cretaceous, Maastrichtian (72–66 million years ago).

CHARACTERISTICS Opal-like in composition, phytoliths are microscopic silica bodies that form in the cells of many plants. The stony particles reinforce plant tissues and slow down predators. An assortment of phytoliths was discovered in dinosaur coprolites (fossil poop), including polyhedral and honeycomb shapes produced by many flowering plants, large pieces associated with conifers, globular, spiny forms produced by palms, and a variety of grass forms.

FOUND IN Dinosaur coprolites from the Lameta Formation.

HABITAT Coprolites were found in lake sediments and associated with bones of titanosaurid dinosaurs. Since these enormous, long-necked dinosaurs could travel great distances, the habitats in which they fed remain a mystery.

NOTES The grasses in this phytolith assemblage demonstrate that the lineage was diverse long before it became ecologically important in the Cenozoic. In addition to grass phytoliths, the coprolites preserved the remains of conifers, cycads, and algae, along with bacteria and fungi that probably helped the dinosaurs digest their food.

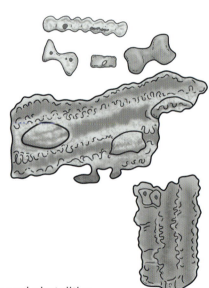

Plant surface tissue and phytoliths found in Late Cretaceous dinosaur coprolites.

FURTHER READING

Ash, S. R. 1976. "Occurrence of the Controversial Plant Fossil *Sanmiguelia* in the Upper Triassic of Texas." *Journal of Paleontology* 50 (5): 799–804.

Bamford, M. K., and I. R. Stevenson. 2002. "A Submerged Late Cretaceous Podocarpaceous Forest, West Coast, South Africa: Research Letter." *South African Journal of Science* 98 (3): 181–85.

Beck, C. B. 1976. *Origin and Early Evolution of Angiosperms.* New York: Columbia University Press.

Beck, C. B. 1988. *Origin and Evolution of Gymnosperms.* New York: Columbia University Press.

Brenner, G. J. 1996. "Evidence for the Earliest Stage of Angiosperm Pollen Evolution: A Paleoequatorial Section from Israel." In *Flowering Plant Origin, Evolution & Phylogeny*, edited by D. W. Taylor and L. J. Hickey, 91–115. Boston, MA: Springer US.

Crane, P. R. 2019. "An Evolutionary and Cultural Biography of *Ginkgo*." *Plants, People, Planet* 1 (1): 32–37.

Dilcher, D. L., and P. R. Crane. 1984. "*Archaeanthus*: An Early Angiosperm from the Cenomanian of the Western Interior of North America." *Annals of the Missouri Botanical Garden* 71 (2): 351–83.

Doyle, J. A., and G. R. Upchurch. 2014. "Angiosperm Clades in the Potomac Group: What Have We Learned since 1977?" *Bulletin of the Peabody Museum of Natural History* 55 (2): 111–34.

Feild, T. S., and N. C. Arens. 2005. "Form, Function and Environments of the Early Angiosperms: Merging Extant Phylogeny and Ecophysiology with Fossils." *New Phytologist* 166 (2): 383–408.

Friis, E. M., P. R. Crane, and K. R. Pedersen. 2011. *Early Flowers and Angiosperm Evolution.* Cambridge, UK: Cambridge University Press.

Harris, E. B., and N. C. Arens. 2016. "A Mid-Cretaceous Angiosperm-Dominated Macroflora from the Cedar Mountain Formation of Utah, USA." *Journal of Paleontology* 90 (4): 640–62.

Harris, T. M. 1940. "*Caytonia*." *Annals of Botany* 4 (16): 713–34.

Herendeen, P. S., and J. E. Skog. 1998. "*Gleichenia chaloneri* – a New Fossil Fern from the Lower Cretaceous (Albian) of England." *International Journal of Plant Sciences* 159 (5): 870–79.

Hermsen, E. J., E. L. Taylor, and T. N. Taylor. 2009. "Morphology and Ecology of the *Antarcticycas* Plant." *Review of Palaeobotany and Palynology* 153 (1): 108–23.

Jagt, J.W.M., M. Deckers, S. K. Donovan, S. Fraaije, R. Goolaerts, R. van der Ham, M. B. Hart, E. A. Jagt-Yazykova, J. van Konijnenburg-van Cittert, and S. Renkens. 2019. "Latest Cretaceous Storm-Generated Sea Grass Accumulations in the Maastrichtian Type Area, the Netherlands – Preliminary Observations." *Proceedings of the Geologists' Association* 130 (5): 590–98.

Johnson, K. R. 1996. "Description of Seven Common Fossil Leaf Species from the Hell Creek Formation (Upper Cretaceous: Upper Maastrichtian), North Dakota, South Dakota, and Montana." *Proceedings of the Denver Museum of Natural History* 3 (12): 1–45.

Judd, E. J., J. E. Tierney, D. J. Lunt, I. P. Montañez, B. T. Huber, S. L. Wing, and P. J. Valdes. 2024. "A 485-Million-Year History of Earth's Surface Temperature." *Science* 385 (6715): eadk3705.

Kelber, K.-P., and J. H. A. van Konijnenburg-van Cittert. 1998. "*Equisetites arenaceus* from the Upper Triassic of Germany with Evidence for Reproductive Strategies." *Review of Palaeobotany and Palynology* 100 (1): 1–26.

Konijnenburg-van Cittert, J. H. A. van, and H. S. Morgans. 1999. *The Jurassic Flora of Yorkshire.* Durham, UK: The Paleontological Association.

Liu, Y.-J., N. C. Arens, and C.-S. Li. 2007. "Range Change in *Metasequoia*: Relationship to Palaeoclimate." *Botanical Journal of the Linnean Society* 154 (1): 115–27.

Looy, C. V., W. A. Brugman, D. L. Dilcher, and H. Visscher. 1999. "The Delayed Resurgence of Equatorial Forests after the Permian-Triassic Ecologic Crisis." *Proceedings of the National Academy of Sciences* 96 (24): 13857–62.

Lupia, R., S. Lidgard, and P. R. Crane. 1999. "Comparing Palynological Abundance and Diversity: Implications for Biotic Replacement during the Cretaceous Angiosperm Radiation." *Paleobiology* 25 (3): 305–40.

McElwain, J. C., and S. W. Punyasena. 2007. "Mass Extinction Events and the Plant Fossil Record." *Trends in Ecology & Evolution* 22 (10): 548–57.

Pigg, K. B. 1992. "Evolution of Isoetalean Lycopsids." *Annals of the Missouri Botanical Garden* 79 (3): 589–612.

Piperno, D. R., and H.-D. Sues. 2005. "Dinosaurs Dined on Grass." *Science* 310 (5751): 1126–28.

Retallack, G. J. 2001. "A 300-Million-Year Record of Atmospheric Carbon Dioxide from Fossil Plant Cuticles." *Nature* 411 (6835): 287–90.

Samylina, V. A., and A. I. Kiritchkova. 1993. "The Genus *Czekanowskia* Heer: Principles of Systematics, Range in Space and Time." *Review of Palaeobotany and Palynology* 79 (3): 271–84.

Schneider, H., and P. Kenrick. 2001. "An Early Cretaceous Root-Climbing Epiphyte (Lindsaeaceae) and its Significance for Calibrating the Diversification of Polypodiaceous Ferns." *Review of Palaeobotany and Palynology* 115 (1): 33–41.

Schneider, H., A. R. Schmidt, and J. Heinrichs. 2016. "Burmese Amber Fossils Bridge the Gap in the Cretaceous Record of Polypod Ferns." *Perspectives in Plant Ecology, Evolution and Systematics* 18 (February):70–78.

Shi, G., A. B. Leslie, P. S. Herendeen, N. Ichinnorov, M. Takahashi, P. Knopf, and P. R. Crane. 2014. "Whole-Plant Reconstruction and Phylogenetic Relationships of *Elatides zhoui* sp. nov. (Cupressaceae) from the Early Cretaceous of Mongolia." *International Journal of Plant Sciences* 175 (8): 911–30.

Srivastava, S. C., and J. Banerji. 2001. "*Pentoxylon* Plant: A Reconstruction and Interpretation." *Plant Cell Biology and Development (Szeged)* 13:11–18.

Stewart, W. N., and G. W. Rothwell. 1993. *Paleobotany and the Evolution of Plants.* 2nd ed. Cambridge, UK: Cambridge University Press.

Stockey, R. A., and G. W. Rothwell. 2003. "Anatomically Preserved *Williamsonia* (Williamsoniaceae): Evidence for Bennettitalean Reproduction in the Late Cretaceous of Western North America." *International Journal of Plant Sciences* 164 (2): 251–62.

Sun, G., D. L. Dilcher, S. Zheng, and Z. Zhou. 1998. "In Search of the First Flower: A Jurassic Angiosperm, *Archaefructus*, from Northeast China." *Science* 282 (5394): 1692–95.

Taylor, D. W., and L. J. Hickey. 1996. *Flowering Plant Origin, Evolution & Phylogeny.* New York: Chapman & Hall.

Taylor, T. N., E. L. Taylor, and M. Krings. 2009. *Paleobotany: The Biology and Evolution of Fossil Plants.* 2nd ed. New York: Academic Press.

Toon, A., L. I. Terry, W. Tang, G. H. Walter, and L. G. Cook. 2020. "Insect Pollination of Cycads." *Austral Ecology* 45 (8): 1033–58.

Willis, K. J., and J. C. McElwain. 2014. *The Evolution of Plants.* 2nd ed. Oxford, UK: Oxford University Press.

Yu, J., J. Broutin, Q. Huang, and L. Grauvogel-Stamm. 2010. "*Annalepis*, a Pioneering Lycopsid Genus in the Recovery of the Triassic Land Flora in South China." *Comptes Rendus Palevol* 9 (8): 479–86.

Zhou, Z.-Y. 2009. "An Overview of Fossil Ginkgoales." *Palaeoworld* 18 (1): 1–22.

INDEX